VETERANS' STORIES Book III:

The Life and Times

David E. Pressey
Author and Editor

Veterans' Stories Book III:

The Life and Times

Copyright © 2015 by David Pressey

All rights reserved. This book, or parts thereof, may not be reproduced or transmitted in any form or by any means without written permission of the author except for the inclusion of brief quotations in a review.

ISBN: 978-0-9968019-1-1 Paperback Edition
         978-0-9968019-0-4 Digital Edition

Library of Congress Control Number: 2015951362

Front Cover Credits: Sergeant First Class Thomas Flowers with his 105mm Howitzer Gun Crew in position during the Korean War, 1950-1951

DISCLAIMER

This book was created to help people understand what it means to serve our nation as a member of the Armed Forces. It was not designed to glorify war or the individuals. Great care was taken to ensure the accuracy of each story and to relate the individual's experiences with the real history at the time they served. Whether the individuals served in actual combat is not an issue since every veteran is subject to the hazards of war and dangers inherent in all military service.

Each individual story tells the roots of the service person and his or her actual experiences during service, followed by post-service life and accomplishments. Some stories are about men who served one enlistment while others are about career men such as Julio Luna, Les Allen, and Thomas Flowers. Thomas Ross is still on active duty as a career Seabee. Edward Kachadoorian is deceased and his story is from family records.

The author shall have neither liability nor responsibility to any person or entity with respect to any loss or damage caused, or alleged to be caused, directly or indirectly, by the information contained in this book.

Printed in the United States of America

DEDICATION

Veterans' Stories –Book III is dedicated to all veterans, active-duty service men and women, and all the members and former members of military reserve components of the United States military establishment. It is in recognition of the individuals who have sworn to give their lives, limbs, freedom, and have risked all future happiness in the cause of the greater common good.

It is especially dedicated to my Army Chaplain, Captain Robert Crane, who was killed in action in March of 1952 in North Korea while serving and ministering to the soldiers of the 160th Infantry Regimental Combat Team of the 40th Division. He talked the talk and walked the walk in the name of Christ.

FOREWORD

This is the third book of Veterans' Stories that are taken from the lives and times of real ordinary soldiers, sailors, Marines, and airmen. The material is organized around the historical events in their early civilian lives: the conditions and surroundings of their pre-service lives.

Their military service is recorded in their own language as they were interviewed or as they wrote their own stories. For older veterans, their post-military service in the workplace and their personal achievements are noted as among the objectives of these books. This information is designed to help the nonveteran public understand that real veterans who have served honorably are among the finest citizens in our local communities.

Though these stories tell of specific veterans and their communities, every community in America has comparable stories based on the personal histories of its veterans.

One aspect of military service is that when a citizen takes the oath of allegiance to uphold this nation and the Constitution and defend it against all enemies, foreign and domestic, then that veteran has given the United States Government the lawful authority to expend his or her life, limb, and liberty as necessary for the "common good" of our wider citizenship and the freedoms that we all enjoy. That a veteran has not had to pay the ultimate price is by the grace of God, for any veteran may be called upon to make the ultimate sacrifice.

This edition of veteran stories has fewer articles than previous publications, but delves more deeply into the roots and backgrounds of the listed individuals and the attendant historical events and antecedents in the veterans' lifetimes. Interspersed among the biographies are articles from military services and experiences common to many different branches of the military.

Table of Contents

Chapter 1	THE LAST CHAPLAIN TO DIE IN KOREA The Story of Captain Crane, Army Chaplain *1*
Chapter 2	COLD, GENERALS JANUARY AND FEBRUARY The Story of Dave Pressey *15*
Chapter 3	THE MOST BASIC OF WARRIORS, THE INFANTRY The Story of Pete Conforti in World War II – 43rd Division *25*
Chapter 4	COFFEE GOES TO WAR, A MATTER OF LIFE AND DEATH *47*
Chapter 5	SECOND COMBAT INFANTRYMAN'S BADGE The Story of Bill Gates, Infantryman *51*
Chapter 6	INFANTRY EQUIPMENT IN WW II AND KOREA *63*
Chapter 7	A MEDIC IN CHINA, A WAR AGAINST DISEASE The Story of Jack Schultz, Army Air Corps India, Burma, and China *69*
Chapter 8	FROM REFUGEE FAMILY TO PATRIOT AIRMAN The Story of Edward Kachadoorian, Army Air Corps *79*
Chapter 9	AMPHIBIOUS OPERATIONS—WW II & Korea *99*
Chapter 10	THE FLOATING DRYDOCKS OF WORLD WAR II Boatswain 2ND Class Orville Edwards *105*

Chapter 11	A CONSCIENTIOUS OBJECTOR GOES TO WAR The Story of Paul Tenbrink *111*
Chapter 12	THE PREOCCUPIED ARMIES OF OCCUPATION Health and Moral Issues of the Army of Occupation *117*
Chapter 13	THE THUNDERBIRD FROM OKLAHOMA The Story of Thomas Flowers, Army Artillery NCO/ Naval Officer *129*
Chapter 14	ATOMIC GUINEA PIGS, CONVENIENCE OF THE GOVERNMENT The experience of Corporal John Pressey RA 28107103 during the Atomic Bomb Tests – November, 1951- Yucca Flats, Nevada *145*
Chapter 15	A LIFE INTERRUPTED — A DRAFTEE'S DILEMMA The Story of Edwin Marks, Artilleryman *151*
Chapter 16	A SCREAMING EAGLE IN VIETNAM The Story of "Doojie" Seliger *167*
Chapter 17	THE BACKBONE OF NAVAL OPERATIONS—THE AIRCRAFT CARRIERS The Story of Commander Les Allen *183*
Chapter 18	FROM UNDOCUMENTED IMMIGRANT TO PATRIOT The Story of Julio Luna, Career Seabee *193*
Chapter 19	THOMAS ROSS—A SERVICE BRAT AT WAR IN AFGHANISTAN The Story of Thomas Ross, Career Seabee *207*

BIBLIOGRAPHY   *229*

THE LAST CHAPLAIN TO DIE IN KOREA

■ ■ ■ ■ ■ ■ ■ ■ ■ ■ ■

The Story of Chaplain Crane, Army Chaplain

Chaplain Robert Crane was my Episcopal Chaplain before all the changes in the Episcopal Church. A member of the California National Guard when he was activated for duty during the Korean War, he was a man of humor, honor, and devotion.

For many of the young men in our combat team, religion had little relevance as they pursued their pleasures, confident in their youth. To most of us, Chaplain Crane was an "old guy," of about 35.

One of our first encounters with the Chaplain was when the troops were receiving their shots from inexperienced medics, who were lined up three deep and on each side of the newly activated Guardsmen. The young Guardsmen were receiving poorly directed shots from these medics, who learned their trade with on-the-job training, using new recruits as their training material. Chaplain Crane was seen pulling a little red wagon with a huge make-believe syringe. Needless to say, some soldiers passed out from the poorly administered shots. They certainly didn't want a reminder of their anticipated discomfort. But it was all in fun.

CAMP COOKE, A RELIC FROM WORLD WAR II

They say that the National Guard was assigned to Camp Cooke, a dilapidated old army base, because it was the cheapest base where the newly activated Guardsmen could be housed. It was dusty and dirty and filled with cobwebs in every barracks. Rattlesnakes lay hidden in the overgrown brush and fields. The base was in an awful state of disrepair. But the Guardsmen, many of whom were veterans from World War II, brought a different mentality to addressing the physical problems of their new home. They trained. They cleaned. They policed the grounds and parade fields. They repaired and reconstructed wherever they were able. Soon, Camp Cooke was a spit-and-polish major infantry, artillery, and armored base.

EVEN THE CHAPELS WERE REPAIRED

As a young seventeen-year-old-recruit with an Episcopalian background, I attended chapel whenever possible. I became more intimately acquainted with the chapel when assigned to a cleaning detail under the direction of a WW II corporal. Duty details went with any and all army training in those days. Sometimes it was KP (kitchen police) a most distasteful experience of washing, peeling potatoes, and cleaning the grease traps. Other times it was barrack detail that wasn't so bad or Charge of Quarters in the Command Post.

Corporal Cadena was in charge of me as the other half of his two-man chapel detail. We arrived at the chapel in the early morning to dust, mop, and sweep the building. Corporal Cadena disappeared until about five o'clock that afternoon, when he suddenly reappeared. I was a little disturbed that he had left me to do all the work by myself. He, however, seemed quite rested and refreshed. It turned out that he'd crawled under the altar and slept all day while I worked. But, rank has its privileges, even if a lowly corporal is in charge of a single recruit.

THE LIMITED MISSION OF THE UNIT

Guard units came much more in the political spotlight when they were transferred from state authority to Federal authority. The mission of the Guard was simply to defend the continental United States against any Cold War Soviet threat while the regular army fought in the bloody war raging in Korea at that time. There was resentment among regular army soldiers that Guardsmen were treated as if immune from the hazards of real war. However, our senior officers and NCOs were experienced soldiers and Marines from World War II. They were not callous youth like the teenage Guardsmen of lower rank and status.

On my 18th birthday, February 28, 1951, twenty thousand men of the Infantry Division were called into a massive formation to hear our Commanding General tell us to forget about the assumption that we were merely a protective force for the Continental United States. He assured us that we would fight in Korea. He warned us to take our training seriously. We knew then that our lives would be in jeopardy.

CRUSTY MILITARY ATTITUDES CONFLICT

I was brought up with three other siblings in a one-parent home by my father. He was a solid student of the scriptures and demanded rigorous adherence to the religious, moral, and ethical codes of the Bible. We were even taught to follow the dietary laws of Moses, and we refrained from eating pork. We weren't allowed to take God's name in vain. Sex was forbidden outside of marriage. Like many of the young soldiers, we did not drink or smoke.

For the old army, however, colorful and profane language was about all our crusty senior sergeants knew. Their vocabularies were limited to the coarsest of expressions. Impressionable young troops mimicked the language and adopted the behaviors of the "old army'" with its earthy approach to discipline and human nature. It was a struggle to keep my morals. It held me back from promotions as the higher ups looked at the strength of my convictions as a weakness. That, however, was the way I was trained, and I honored my father by following his edicts.

THE FORMER MARINE TRIES TO SHIELD ME

Our 1st Sergeant, a former WW II Marine named Torpey, looked at me as a "fish out of water." He discovered I had joined the National Guard illegally one day short of my 17th birthday. He called me into the command post and told me he was going to discharge me because my father had not signed for me and I was underage. I begged him not to send me home. He was quite forceful in telling me that I didn't know what war was all about. He emphasized the hazards of combat. But he never threatened to discharge me again, and the issue of being underage and illegally in the army was never brought up again. Instead, he took a new tack: he tried to find and then assign me to every relatively safe position in the rifle company. He tried to get me to be a cook. I told him I just wanted to be an Infantry soldier like my father. He tried to make me a supply clerk. I told him the same thing. I told him I liked the army, which he thought was odd. He was a good man and I knew he was trying to look out for me.

Since I followed the Biblical dietary laws, he thought I might be Jewish. I told him that was the teaching from my father and that I was an Episcopalian who followed the Bible. That sure confused him, but he ordered the cooks to give me extra protein food such as cheese and eggs to make up for the ham and bacon so common in military mess halls. He even talked to the Rabbi Chaplains to see if there was a dispensation that would allow me to eat regular Army food. They acknowledged that there was a dispensation related to Kosher Laws. The 1st Sergeant brought that to my attention. I explained that I would follow what my father taught me and that I was a Christian baptized and confirmed in the Episcopal Church. The sergeant inquired of the Episcopal chaplain if that was a tenant of the Episcopal Church. The Episcopal Chaplain stated that that was not part of church doctrine and teaching.

CHAPLAIN'S ASSISTANT—THE PERFECT ANSWER FOR THE SERGEANT

One day 1st Sergeant Torpey had a solution that he thought would remove me from all the hazards, pain, and discomforts of the Infantryman. He called me to the Command Post and told me that he had the perfect assignment for me: I was to become Chaplain Crane's assistant.

I asked, "What's his assistant do?"

The 1st Sergeant said that I would shine the chaplain's boots, tidy his living quarters, and hang up his clothes and vestments. I would take care of the religious accoutrements and assist in the services as an acolyte. Most of all, I wouldn't have to be out in the muck and mud of the field. He said, "It's far better than being an Infantry rifleman."

I said, "I don't want the job. I just want to be an Infantryman like my father and like my brother." My brother had just been wounded in the fighting in Korea. I said, "I don't want to be a butler."

At that the sergeant threw up his hands, muttering that I was too stupid to know what was good for me.

Little did any of us know what lay in the future, for Chaplain Crane was not a laid back Chaplain. He was a soldier's Chaplain—one of the "old school."

A soldier listening in the back of the Command Post piped up, "If Pressey doesn't want the job, can I have it?"

In frustration, the Sergeant said, "Yeah, Pressey's just too stupid to know what's good for him. Sure, it's yours."

NEW MISSION FOR THE GUARD

Despite the so-called political shield protecting the National Guardsmen from combat and overseas deployment, my Infantry Division found ourselves on our way to Occupied Japan with a new mission—to protect Japan from Soviet intervention—coupled with rigorous combat training for any possible contingency. The training intensified on the slopes of Mt. Fujiyama, where training ratcheted up to full-scale, all-live-fire, regimental-assault coordinated air strikes, artillery firings, and tank shellings. The training was so intense that we experienced casualties. We refined our Infantry skills, and I became a machine gun squad leader.

SOLDIER TOWNS: CENTERS FOR VICE

Towns adjacent to military camps frequently become centers for vice and sordid entertainment. They are not Disneyland. Young soldiers are easy pickings for the denizens of these establishments of iniquity. With the excitement of freedom from moral restraints, even the most disciplined moral soldiers can succumb to the temptations. Drinking, smoking, prostitution and other vices abound in the towns near military bases.

Vices and immoral behaviors in Asia exceeded anything in the States, since there were no religious, family, or moral constraints on the soldiers. Religion and religious services almost became extinct.

At Camp Mc Nair on the slopes of Mt. Fujiyama, Sunday services were conducted in a 10-man squad tent with a dirt floor. Chaplain Crane set up his altar and performed his services of Holy Communion as if he were in a large cathedral. He didn't neglect one rubric or shorten one sermon. Yet, out of 5000 men in the Regimental Combat Team, he seldom had more than three to five soldiers in attendance. Sometimes Chaplain Crane needed help setting up the altar and accouterments used in the service.

One Sunday morning as I entered the tent, he asked me to get some Holy Water. I was perplexed and said I didn't know where to get such an item. What does a kid know about Holy Water? He pointed across the dirt street to a Quonset hut and directed me to a water spigot. He said, "That's where you get Holy Water." I had thought Holy water was some kind of special water or substance derived from some sacerdotal act or maybe from the Holy Land. Chaplain Crane was always practical and direct. No hocus pocus and no magic!

Unbridled, licentious behavior began to take its toll with the spread of venereal disease in our camp. It was beginning to impact the training and hence our mission to defend the Japanese Islands from the Soviet threat while war raged in Korea. The Chaplain Corps was called to give moral talks appealing to the troop's faith and moral training. They appealed to the sense of responsibility to soldiers' wives, children, or their girlfriends at home. The Chaplains tried their best to remind them of faith and virtue. But temptations were too great and the chaplains were usually ignored! Their talks did little good. Boys will be boys, especially when no one will be around to monitor their behavior. There was too much fun in Yamanaka to be intimidated by a few army chaplains!

The army isn't concerned about morals and manners except when they interfere with the mission. Disease was becoming a major threat to the fighting efficiency and training program. The army is always practical in its approach to problems, so they took a different tack. Using medical knowledge about disease control, they began to educate the soldiers in platoon-size classes about the effect and control of infections. Every soldier was required to receive a prophylactic kit before he could go on pass to Yamanaka and the geisha houses. Demonstrations were given how to put on a condom for protection. During one demonstration, a chubby Texas-Mexican soldier was asked to actually show the platoon how to put on the condom. The troops were shocked as Serrano unbuttoned his fly and reached into his pants for his organ of copulation. He reached deep as the stunned troops watched. Suddenly, he pulled out – the nozzle of a gas can, which he promptly used for his demonstration. Needless to say, by this time the troops were rolling in unrestrained laughter. But the lesson registered. The medics gave talks on proper follow-up after an encounter at the dispensary where they gave shots of penicillin.

A WILD CHRISTMAS PARTY IN THE CAMP

The rifle company leadership decided to sponsor a Christmas party for the men of Fox Company using funds from the enlisted men's special account. All the officers of the regiment were invited. The party was to be held in a huge steel-framed building used as a mess hall, the only permanent structure, along with the dispensary, in the 500-tent encampment.

Most of the officers planned to attend except for Chaplain Crane and the Catholic Chaplain. They issued a statement: "We do not celebrate the birth of Jesus Christ in drunkenness. We will not attend." The other chaplains planned to attend, and they did. The reaction of the troops was mixed. Some men respected their stand. Others thought they were "killjoys" and were not being one with the troops.

Being a nondrinker and a "fish out of water," I hadn't planned to attend either. However, our new 1st Sergeant called me to the Command Post and told me that I was assigned to be the bartender in charge of serving the officers. I would have two KPs to assist me. Since I was underage, that would not be allowed in the States. I protested. I said, "Why me? I don't drink."

The sergeant replied, "That's why you're in charge. You'll be sober when everyone else is drunk."

As Chaplain Crane had foreseen, the party was primarily a drinking affair with an assortment of so-called Geishas in attendance. The enlisted men were separated from the officers and were served plenty of beer, while the officers had their choice of beer or a gin concoction ladled out of huge aluminum cooking kettles. As the night wore on, I noticed my KPs would serve an officer a drink out of the gin kettles and they would take a drink themselves. Soon the serving became a weaving dance between the officers and the KPs, since they were all drunk as the evening wore on. Most appalling to me was the drunken chaplains who engaged in this bacchanal ritual dance of inebriation. I watched as the KPs poured liquor all over the sleeves of a drunken officer's "pink" uniform. The first sergeant was right! Few KPs were left standing after the end of the "Christmas party."

My two KPs were unconscious on the floor before the night was over. With a skeleton crew of those left standing, we worked to the wee hours of the morning to clean the mess hall.

THE REVIVAL OF WORSHIP AND FAITH

Many of the men of the regiment believed we would never enter combat in Korea because of our special status as a National Guard Division. I continued my training on Mount Fujiyama at Ski Trooper School. Suddenly, all training ceased. We were called to formation where it was announced that we were to report to our rifle companies now at Camp Zama near Tokyo. It was announced that we were going to the war in Korea, where we would relieve one of the army divisions on the front line. The assumed safety of being a member of a politically sensitive National Guard Division ended right then and there! Suddenly, a quiet fear settled over the men as they began to rethink their mortality and the possibility that they might be held accountable by their Maker.

There was a surge in religion, religious service attendance, and the revisiting of prayer books and Bibles. One soldier even purchased a special Bible with a supposedly bulletproof cover that he carried in his left breast pocket. He was hoping that that might save him. Catholics went to confession and mass. Protestants returned to their Bibles. I'm sure Chaplain Crane was quite busy at that time, not only with Episcopalians but with anyone with a spiritual need.

Military chaplains are very ecumenical, and serve the needs of all soldiers of all faiths.

CHAPLAINS IN THE TRENCHES

Our regiment entered combat operations up in the Taebek Mountainous spine of North Korea. The war had devolved into World-War-I-like trench warfare with constant artillery and mortar fire. Combat patrols were active down in the valleys, where their movements were punctuated by firefights and ambushes. Connecting trenches tied the machine gun bunkers, foxholes, and fortified positions into one continuous line that stretched across the Korean Peninsula. Like animals, the young soldiers lived in frozen holes and bunkers. In some places they were surrounded by the bodies of dead enemy soldiers in a frozen hell. It was shear misery just trying to keep from freezing to death in that environment, especially when any careless exposure would ensure that you would be targeted by a sniper and/or enemy mortar or artillery fire.

Chaplains are not combatants. They do not carry weapons; yet they were in the thick of the battle area where they came on line to minister to the same young soldiers who had ignored the chaplains when back in Japan. The chaplains were a liaison between anxious parents and families who often did not hear from their loved ones who were now in mortal combat.

Chaplain Crane was no exception, for Episcopal Chaplains have been men of courage and dedication throughout history. It was while entering the combat zone to minister to the suffering men and boys holding the front that Chaplain Crane was targeted by enemy artillery or mortar fire. In the barrage that followed, the good Chaplain was killed, and his assistant, the one who thought he had a soft cushy job where he would be safe, was severely wounded.

The men of our regiment were shocked and heartsick that such a noble man should die. Those who had made remarks about his absence from the drunken Christmas party in Japan now remembered him as a man who "Talked the talk and walked the walk."

Several weeks later, memorial services were held throughout the front line by the other chaplains who met with small groups of soldiers, even on outposts miles in front of the main line of resistance and deep in no-man's territory, surrounded by hostile forces.

A DANGEROUS SPOT FOR A MEMORIAL SERVICE

As members of Fox Company, we were on a platoon outpost miles in front of the lines. There were about 45 of us looking down on an enemy-held rail center in Kumsong, North Korea. A chaplain had come up to that area. He located himself in a draw a short distance outside the barbed wire surrounding the outpost.

Small groups of soldiers went outside the outpost to the sheltered draws for the memorial service. Their weapons in hand were locked and loaded, and each man carried two grenades. Browning automatic-weapons men ensconced themselves above the draw in a defensive position while below, the chaplain set up his little altar stand and preached a memorial homily honoring Chaplain Crane and all he represented. It was a short but touching ceremony, and we nervously kept our weapons ready as we kept looking around for any unwelcomed enemy who might approach.

CHAPLAIN CRANE LIVES ON

By his faith and his uncompromising witness to God's truth in humble service, Captain Crane became an example to all who knew him. He was only about 35 when he died, but he was an old man by the standards of young combat troops.

Over time, most of the young soldiers returned to the States and became civilians again. They reverted to the training of their youth and repented any misdeeds as they reconnected to the values of faith and family, those very values always upheld by my Chaplain.

COLD, GENERALS JANUARY AND FEBRUARY

■ ■ ■ ■ ■ ■ ■ ■ ■ ■

The Story of Dave Pressey

BATTLES IN INHOSPITABLE PLACES

The carnage of the battlefield is what many people believe to be the main cause of human destruction, but more war-related deaths and injuries are caused by weather and disease. It's rare that military planners select balmy and pleasant environments for warfare, particularly in modern times. In more medieval times, armies waited until spring to commence combat operations. There are some exceptions. The Mongolian Hordes of Genghis Khan often transited Asia and Eastern Europe during snowy conditions, to which the Mongols and their warhorses were accustomed. George Washington took a recess at Valley Forge during the harsh winter months except for the one winter attack on the Hessians at Princeton. That was unusual, and not expected by the Hessians. Russia has had two great generals who always defeated invading armies from the West. Their names are Generals January and February. Cold winter weather defeated the armies of Napoleon and, later, the German army. Cold is a formidable adversary of any army.

CALIFORNIANS LACK COLD WEATHER EXPERIENCE

Southern California lads looked upon snow as a unique, fun-filled experience. It's not uncommon for Angelinos to look up at the San Gabriel Mountains and Mt. Baldy and the deep snows of the high mountains while strolling in shorts and shirtsleeves in comfortable 80-degree weather in the valleys below.

California snow is usually wet and mushy compared to the dry powdery snow of colder climates. If one becomes too chilled playing in the Southern Californian snow, a quick trip down the mountain is all that is necessary to warm up. Southern Californians don't understand or appreciate cold. Frostbite is an unreal abstraction. Cold is discomfort but rarely a condition that threatens life or limb.

My father, who was from Vermont, rarely talked about his winter experiences in that cold state where winter is almost six months of the year. He did talk about the summers in glowing terms, but whenever I suggested that we move to Vermont, he would suddenly have the most pained look on his face. He explained that he couldn't

COLD, GENERALS JANUARY AND FEBRUARY | 17

handle the winters. It took Korea for me to understand his attitude toward the long winters and bitter cold of northern climates.

Cold in Korea

FIRST ENCOUNTER WITH LIFE-THREATENING COLD

My first encounter with real cold was in Asia when I was a teenage soldier. We loaded APA Navy troop ships in Yokohama Harbor during a snowstorm. It was a gentle snowfall, and not extremely cold. For five days we sailed south from Yokohama to the Yellow Sea. With hundreds of troops in the holds of the ship, we were relatively warm and comfortable.

In the gray dawn of the fifth day, excited troops came into my compartment saying that we had arrived at Inchon, Korea. I rushed up the ladder (or stairs) to the deck where in the twilight I could just make out the outline of the shore. I grabbed the steel railing. Instantly my hands and fingers began to freeze. I felt pain in my fingertips. My toes began to freeze. I drew back my hands and bits of skin remained frozen to the rail. Even my ears began to freeze. It was colder than anything I had ever experienced. I hurried back to the warm compartment.

A new awareness of cold penetrated my psyche. I began to dress for the cold. First I put on one pair of cotton socks plus two pairs of wool socks—I was taking no chances. I put on a cotton T-shirt and shorts. I pulled on a pair of long johns and a heavy undershirt. I then put on wool trousers and shirt. Over the trousers I pulled on sateen snow pants and a faux-fur-lined jacket. Next, I put on an oversized field jacket. Lastly, I wrapped this layered clothing with a trench coat. I put on a fur-lined cap with earmuffs. Over that, I put on my helmet. On my hands, I put on a pair of wool mittens and over that I added a pair of leather gloves. (Even then my fingers started to freeze when out in the raw elements of the Korean winter.) Lastly, I had a special set of gloves with a trigger finger for firing as needed. My boots were "shoepacks," a leather and canvass monstrosity that failed to keep anybody's feet warm. The mass of clothing and equipment restricted my circulation and movement as I picked up my sleeping bag, rifle, bayonet, canteen, cartridge belt, pack, and helmet.

Eighteen thousand troops off loaded into landing craft. We moved toward shore to plank walkways in the tidal flats of Inchon Harbor. When we reached the shore, there was a short incline, slippery with ice. I slipped on the ice and fell. The weight of all the gear, clothing, and weapons prevented me from regaining my feet. Two buddies walked up to me and jerked me to my feet as we waddled off to war. And we were expected to fight under these circumstances! We felt clumsy and foolish.

We gathered around a roaring bonfire up from the Inchon beach. A field kitchen had been set up nearby to serve hot coffee to the troops. Suddenly, I heard a crack that sounded like a rifle shot. Supposedly, Inchon was secure from enemy activity but the crack startled me until I realized that the water in my canteen, which I had filled to the brim on ship, froze as we landed and the expanding water cracked the canteen lid. I took the cup from my canteen pouch to get some hot coffee from the field kitchen by the bonfires. When the coffee was poured into the cup, I accidently spilled it on my layered uniform. I was a little distressed to soil my uniform the first day in Korea. As I went to wipe off the spilled coffee, it had frozen before penetrating the fabric. I merely peeled off the ice and brushed it away. By then, I was cold, real cold! Mentally, I was unprepared for such cold. I was a Southern California lad, but it began to dawn on me that what we'd learned in winter warfare classes in Japan was serious stuff.

I remembered the admonishments: "Lose a glove, lose a hand. Freeze a foot, lose a foot." Reality hit hard. Cold is not a snow trip to the San Gabriel Mountains in California. Cold is the main enemy of all soldiers living in the foxholes and trenches of North Korea. Cold could maim! Cold could kill! For the rest of the winter of 1951-52, cold was the enemy we could never ignore. But cold wasn't just *our* enemy; cold was the enemy of the communist soldiers as well. They suffered even more with their quilted-cotton uniforms, wrap-around leggings, and tennis-type shoes.

KOREAN WINTER IN THE TRENCHES

From Inchon, we boarded two-and-a-half-ton trucks for transport to the far northern reaches of the battlefront in the Taebaek Mountains. Boarding the trucks gave some relief from the biting wind chill of Inchon Harbor unless you were unlucky enough to sit on the open end of the canvas-covered trucks. In the interior of the truck, we had some comfort from the body heat of soldiers huddled together on the rough and bumpy ride. The importance of conserving body heat reminded us of other instructions from our winter warfare training.

The convoys of the 160th Infantry Regimental Combat Team moved across the frozen landscape of the cold and barren flats between Inchon and Seoul. Trucks and heavy tanks clattered across frozen rivers. Can you imagine a 32-ton tank or self-propelled artillery piece going across a frozen river without breaking the ice? But the ice was thick enough to support them. That's how cold it was.

The color of cold was gray and white, except for the dark green forests, as we began our motor march high up into the mountains—where it was even colder. Another lesson: cold increases with elevation, and also as the distance between us and Siberia was shortened.

Eventually, the trucks pulled into a "valley of fire," not from heat, but where massed artillery was almost hubcap to hubcap, interspersed with tents, piles of ammunition, and bunkers. The tents were in front of the firing artillery because the valleys were so deep and narrow that the artillery had to back against the opposite side of the valley so the rounds could clear the hill in front of them.

We drew C-rations and were given naphtha pills for heating the contents of the cans of food. Normally, one tablet would heat one can—that is if we were in Hawaii or California. But in Korea's winter, a week's supply of naptha pills could be used for just one can: burning the bottom, thawing the middle, but keeping the top semi-frozen. We soon learned that by carrying the C-rations under our clothing, we could help thaw them. Jeep and truck drivers could heat them on an engine, but if they became too hot, they could explode.

There were other problems caused by cold. Water-cooled machine guns had to have coolants that wouldn't freeze. Special attention was required for motor vehicles: for oils and greases. Weapons couldn't be oiled because congealed oil might cause malfunctions. Sanitary needs couldn't be remedied on the front lines with the normal burial of waste. Digging a "cathole" was impossible in the concrete-like frozen earth. Human waste in the snow could cause real sanitation problems when the snow began to melt. Fortunately, the bodies of dead enemy soldiers weren't a health problem until the thaw of springtime.

THE DEMORALIZING EFFECTS OF CONSTANT COLD

City boys don't read weather. They have to experience it in the raw, often to their regret. A clear, still, starlit night can send the mercury way below zero Fahrenheit. To stand guard in a foxhole machine gun nest is sheer misery in such weather. The soldier is in a continual war against the elements. Constant movement of the fingers and toes is necessary to maintain circulation and prevent freezing the extremities. Once frozen, the living cells are killed, and when they thaw, the dead cells become gangrenous and require amputation.

Patrols are particularly hazardous as they move to positions beyond the main line of resistance, lying in wait to ambush enemy patrols. Stealth and stillness are required to avoid detection. Numbness in the extremities and frostbite are the result of enduring a night out in the open. Yet, despite the constant struggle to keep from freezing to death or suffering frostbite, cold is endurable. It was endured by 1.7 million Americans who served in the Korean War.

The worst cold is when wind combines with snow flurries, producing a wind chill that causes an intense agony of mind and body. It's easy to understand how the Norsemen suffered and went crazy during the long, dark, freezing nights of their northern countries. They had a word for the mental breakdown under such circumstances. They coined the word "berserk" to describe the frenzied state that could occur under adverse conditions of cold and darkness of winter. In those blizzard conditions, the face, whiskers, hair, and eyebrows of the unshaven G.I.s are encrusted with snow and ice.

Eighth Army orders directed the troops not to use earflaps while standing guard in those awful blizzards, or to build fires. Rear area directives ordered all soldiers to face the front with their ears uncovered so we could listen and react. Survival dictated that we ignore such orders from commanders who were immune to the realities of the troops on the line, troops who were simply trying to survive the bitter cold—for cold was often more of a hazard than enemy troops. We knew that the enemy was suffering even more than we were because of inferior clothing and equipment. We often said to each other, if an enemy soldier could find his way up the slopes to our positions in a raging blizzard, they were better men than we. We knew our adversaries were human. We'd seen their frozen bodies in their pitiful quilted-cotton uniforms, wool leggings, and tennis shoes. They suffered worse than we did. Yes, the troops who endured the Korean winter, and for that matter, all troops who had to fight and endure bitter cold, know what it means to endure, not for a moment, but for day after day, week after week, and month after month, until the spring thaws begin.

PERMANENTLY CURED OF PLAYING IN THE SNOW

I have felt cold. I have heard the howling blizzards with cold in my face. I have lived with cold such as few experienced it as a front line soldier. Korea cured me forever from any desire to drive up the snow-covered mountains of Southern California. The closest thing to cold that I care to contemplate is a Christmas card with a snow scene. Give me the subtropical climate of the Southern California valleys. No one ever froze to death in such a salubrious climate.

Dante in the *Divine Comedy* writes about different levels of hell. I believe he stated that the lowest level of hell was not hot, it was frozen. The Infantry soldiers of World War I, World War II in Europe, and the soldiers of Korea would most certainly agree.

THE MOST BASIC OF WARRIORS, INFANTRY

The Story of Pete Conforti in World War II – 43rd Division

As told to and written by Mr. Conforti's daughter, Mary Nugent.

I'll start when I was in high school (in Santa Cruz, California). In March of 1943, I was in my senior year when I received a telegram from the War Department saying I was to report for my physical exam. I was eighteen years old and still had three months before my graduation. I contacted the Department and received a deferral until June. I graduated on Friday, the Fourteenth of June, and reported for my physical three days later on Monday the Seventeenth.

A couple friends and I took the bus from Santa Cruz to San Francisco. I remember that the center was really busy with other inductees. After passing our physical, my friends and I decided to go to Oakland and see a girl that I'd dated in Santa Cruz. Her name was Marion Eaton. After seeing her, we decided to stay in Oakland for the night. Not having enough money for a room, we found an all-night theater and slept in the seats. Next day we returned to Santa Cruz. I had a 1936 Ford sedan and had put a new engine in it several months before. I was hoping to sell it before reporting for duty on the first of July and had a couple of offers, but I decided to hold on to the car. I could have taken it with me, but didn't think that was a very good idea.

Pete Conforti,
Camp Roberts, California, 1943

On July 1st, I said goodbye to Mom and my brother Jim. My dad was working in San Francisco at the shipyard. The bus took us over to the Presidio in Monterey, which was the induction center. This is where you receive your uniform and indoctrination. While there, I was also interrogated regarding my nationality. They asked me if I had any qualms about being sent to Europe and fighting Italians. Being an American, that was not an issue for me. I must have spent about one week at the Presidio and finally got orders to report to Camp Roberts, California, for seventeen weeks of Infantry Basic Training.

Camp Roberts is located about fifteen miles north of Paso Robles in the Salinas Valley. Arriving there on July 1st it was quite a shock, as the daytime temperature is always around 100 degrees. The fortunate thing was that the nights were very pleasant. The base was also a training camp for field artillery. I think there were over 50,000 trainees there. I was assigned to the 82nd Infantry Training Battalion.

It was fortunate that I had just graduated from school and was in very good physical condition. In high school during our senior year, we were given extreme physical training during our gym classes. We were doing obstacle courses, hill climbs, rope climbs, push-ups, chin-ups, sit-ups and one-mile runs. All of these events were timed. With this conditioning, the Army training was not too difficult, even while carrying a sixty-pound field pack on hikes.

After a couple of weeks of basic, we were finally able to get twenty-four or forty-two hour passes. The forty-two hour passes were hard to get, and there were only so many given out. I was lucky and was able to go home to Santa Cruz about four or five times during my training. One of the weekends there, I sold my car for 600 dollars. The forty-two hour passes we got were from 12 noon Saturday to 6 AM Monday. When we went home we had to wear our uniform at all times. We couldn't change into civvies.

BASIC TRAINING

Basic training consisted primarily of battlefield tactics. We spent several weeks on the firing range learning to shoot the M1 Garand rifle, 30-caliber machine gun, Browning automatic rifle, 60 mm mortar, and pistol. We also did twenty-mile hikes with overnight bivouacs and forced marches. We were the last class at Roberts to have the seventeen-week course as the next group was cut down to thirteen weeks. This was an indication of the need for more manpower overseas.

I finally finished basic around the first week in November, 1943. We were given one week's leave and then were ordered to report to Fort Ord. I spent about two weeks at Fort Ord and received additional training that included crawling under barbed wire with machine guns firing over our heads. Pretty scary!

With Fort Ord out of the way, I was transported by train to Camp Stoneman. When you went to Stoneman, you knew that the South Pacific was waiting. Camp Stoneman was known as a POE (Point of Embarkation) and was located near the town of Pittsburg, California. Finally, on December 9, 1943 we boarded an old paddle wheel ferry on the Sacramento River and paddled down to San Francisco. There we boarded a troop ship called *General John Pope*. It had been a passenger liner before it was renamed after a Civil War general. On board, I think there were about eight thousand troops. We left San Francisco the following morning and I still remember going under the Golden Gate Bridge.

As soon as we were underway, we were assigned duties. I was assigned to the mess hall. We had twenty-four hours on duty and twenty-four off. This was okay until I started to get seasick about the second day at sea. Food was the last thing I wanted to see so I was assigned to wash pots and pans. After a couple of days, I finally got over the sickness. Working in the kitchen was pretty good, as we got first choice on the food being served.

We found out we were headed for New Caledonia. It was to be about a two-week trip and on the way, we were "blacked out" at night and were going on a zigzag course to evade any Japanese subs. We didn't have any escort ships, so we were on our own. We must have been cruising around eighteen knots. On December 23rd, we docked in Noumea, New Caledonia. The island is owned by France and is located east of Australia. We boarded trucks and were driven inland about twenty miles to a camp that was used as a replacement center for all of the Army divisions fighting in the Solomon Islands.

Two days after arriving there, we celebrated Christmas. I don't remember anything about it. While waiting to be shipped out, we would be assigned to other units located on the island to assist them in any way possible. One that I remember in particular was an all-Negro company that handled the disbursement of motor vehicle parts. While there, we were served "soul food" in their mess hall. I think it was some of the best food I ever had while overseas!

Pete Conforti and Fred Carruthers in New Zealand Encampment

After being there for about a month, several thousand of us were moved to a camp near the town of Noumea. We were getting ready to join the 43rd Infantry Division. We were supposed to join them on New Georgia in the Solomon Islands. While in Noumea, we were able to swim at the beautiful beach near our camp. The water was so clear due to all of the coral. The only bad feature was the great white sharks in the area. We always had someone on guard with a rifle. Another favorite place to visit was the "Pink House."

HIT BY A TYPHOON AND ON TO NEW ZEALAND

While in Noumea, we were hit by a typhoon. The winds were blowing around 100 mph and by morning, all of the tents in camp had been leveled. A lot of wind and rain but as far as I know there were no casualties. Around March, 1944, we were loaded aboard a ship and were sent to New Zealand to join our division. We debarked in Auckland and were sent to our camp about twelve miles south in a small town called Papatoetoe. The camp was right next to the railroad tracks and station. It was here that I was assigned to Company A, 172nd Infantry Regiment. I was assigned to a rifle squad, and slept in a six-man tent. The tents all had wooden raised floors so they were quite comfortable even with the cold season coming on.

The camp was called Cambria Park. New Zealand is a very beautiful country and it resembled California, especially with the rolling hills. The people were very friendly to us servicemen.

NEW ZEALAND, A GOOD EXPERIENCE AND FRIENDLY PEOPLE

At the time, I was receiving sixty dollars per month, at the rate of fifty dollars plus twenty percent for overseas pay. I was able to send about half of this home to Mom and still had enough to go into Auckland for a show and dinner. I did do some short sightseeing trips and used to go down to see Fred Caruthers. He'd ended up in the 118th Engineers, and was stationed about fifty miles further south. Fred and I used to pal around together while attending Santa Cruz High School.

One weekend another solder and I got a two-day pass and took the train to a town about 125 miles south called Hamilton. When we got there, we didn't see any other G.I.s and before we knew it, we were getting offers from the local residents to spend the weekend at their houses. We accepted such an offer. It was nice to eat and sleep in a home.

The autos in New Zealand were all using charcoal to generate the gas required. They looked strange with their burners attached to their bumpers.

TRAINING FOR JUNGLE WARFARE IN THE SNOW!

Around the first of June, we were shipped to a place called Rotorua. This part of New Zealand is full of steam baths and looks pretty hostile. We were starting to pick up some snow. We were practicing for jungle warfare in the snow. While we were there, we heard that the Allies had landed in Normandy. This was very good news for us as it meant the war was progressing favorably. After about two weeks of jungle training, we returned to our camp.

While in New Zealand, I dated a couple of girls from Auckland. One of the girls worked in a soda fountain shop so I was treated to milk shakes. Steak and eggs is a popular dinner meal there so it was not long before I learned to love this meal. We also ate a lot of ice cream and drank gallons of milk. The water in New Zealand was terrible and was the reason many of the people there have bad teeth.

ON TO NEW GUINEA AND ACTIVE COMBAT

Around July, 1944, we received orders to ship out. We boarded an old Dutch freighter that was not equipped to handle any passengers. I think that on board we had the whole battalion of about 800 men. The ship had wooden decks so we would end up sleeping on deck. The kitchen was also set up on deck and we were fed only two meals a day. We were in a convoy so we had US Navy escort ships. We felt pretty safe.

We were headed for New Guinea. We were going to Aitape on the north central coast. This was about a 3500-mile boat ride that took about two weeks. We were to relieve the 32nd Infantry Division. Aitape is on the coastline and planted with coconut trees, owned by the Colgate Palmolive Company. The tree farms were about one quarter mile in width and must have run thirty miles along the beach. Once past the trees you were in the jungle.

While there, we were on patrol duty into the jungle. The Japanese were trying to retake an airfield that we'd captured. On some of our patrols, we had to spend nights in the jungle. The nights could be scary, with all of the noises from birds and the rats as big as cats that would jump around your foxhole at night. Another thing you could depend on was the rain in the late afternoon. You could hear it coming from a far distance. Fortunately, it didn't rain long but when it did, it was so loud you couldn't hear anyone talk.

We were warned about a native village along the Driniumor River. We used to have to patrol near it, and the natives there were headhunters.

Except for the patrol duty, it wasn't too bad while we were in Aitape. Bob Hope and his USO troupe put on a show for us that was very entertaining. Sometimes we were assigned to ships in the harbor to be unloaded. Around September and October, we started an intensive amphibious training program. The rumor spread that we were going to land in the Philippines.

While in Aitape, I started playing softball with our company team and I became the starting pitcher for our regimental team. It was fun because we used to be excused from certain details.

We spent a lot of time on the rifle range. They must have liked my scores as I was selected to be the company sniper. This was the farthest thing from my mind that I would ever do. I was given a 1903 Springfield rifle with a scope. It was brand new and I had to clean the Cosmoline (a grease-like coating to prevent rust) off. With the scope on, I could hit an eighteen-inch bull's eye at six hundred yards. The rifle really had a kick to it compared to the M1 Garand. It was the same gun used by the military in World War I. I never did fire it in combat though. After we landed, it was decided that it was never needed.

We used to get rain almost every day and with the humidity there, your shoes would rot in about three months. Also, if anyone was wounded or had a broken limb they were evacuated to Australia because wounds would not heal in that climate. We were also having a lot of guys go down with malaria. We were forced to take Atabrine tablets to keep from getting malaria. As we went through the chow line, an officer would make us open our mouth and he would flip a pill in. The pill had an extreme bitter taste to it, but it sure seemed to do the trick for the most part.

One of the side effects of taking Atabrine was that it turned your skin yellow. This was not apparent to us as we all looked the same to each other. But it really showed up when we got some replacements and we could see the color of their skin.

Also, while in Aitape I became engaged. Her name was Artie Hornbaker. I'd known her for a couple years in Santa Cruz and used to see her often. She and Chickie Matovich were very good friends. We were corresponding while I was overseas and I always had a crush on Artie. I explained this to her in one of my letters.

ON TO RETAKE THE PHILIPPINES FROM THE JAPANESE

A lot more rumors were going around that we were headed for the Philippines. Around December 28, 1944, we boarded our ship and pulled out of Aitape harbor to join other convoys. Our destination was Lingayen Gulf, with the landing scheduled for January 9, 1945. The ship I was on was called the APA *Du Page* and it was capable of carrying one battalion of about 800 men. The ten-day trip to Lingayen Gulf was uneventful, as we were well protected by navy ships. By the time we arrived there, the landing area had been under fire from battleships and cruisers. The total ships involved were around 800 and it was to be the largest landing in the Pacific. It involved four divisions of 60,000 men.

Our regiment was on the left flank of the landing. As we were unloading our gear, we could hear the tremendous noise coming from the fourteen- and sixteen-inch guns of the battleships. I heard that one of the ships was *California,* which had been damaged at Pearl Harbor.

As we started to climb down the sides of the ships using rope ladders to get into our LCPs (Landing Craft Vehicle/Personnel) it suddenly hit me—this was not an exercise anymore. This was it. The LCPs hold about 40 men. After leaving the ship, the LCP meets other LCPs and we go around in a circle until we are directed to shore. We were about two or three miles from the beach and then we were on our way in as the third wave.

The first wave was to land at 9:30 am so we came in around 9:45. Upon our landing, we came under slight mortar and small arms fire. We proceeded to move rapidly, intending to capture the hills assigned to us.

We made it inland about one mile and secured our first objectives for the day. We hadn't received any casualties, and we prepared for nightfall. We were up high enough to see the Lingayen Gulf, still teeming with ships. Just before dusk, we could see enemy planes attacking the ships. The sky became covered with ack-ack (antiaircraft) fire from the ships. We heard later that the *Du Page* APA troop transport had been struck by a Kamikaze plane that hit the bridge. Around thirty-five sailors were killed. While on the ship I'd met a sailor from Reno. I always wondered if he'd been injured during the Kamikaze attack.

FIRST NIGHT ASHORE AND COMBAT OPERATIONS

The first night we were secure enough that we didn't even dig foxholes. The next day we started to advance in a northeast direction. We still hadn't met any resistance. We came upon our first enemy that had been killed by shells. There were around five or six bodies that had been dead for a while. It was my first glimpse of seeing human bodies all bloated. The stench was awful.

In combat, you lose all sense of what day of the week it is. I'll try to describe some of the events that occurred: The foxholes we dug every night were for two men. They were about five feet by three and one-half feet and about twelve to fifteen inches deep. The type of soil determined the depth. The area to be covered played an important part in the spacing of the hole from your buddies. They could be from ten to twenty feet apart. You always dug in before you ate any dinner. One night we were dug in with one squad (about twelve men) about seventy-five feet in front of the side of a ravine. About midnight, all hell broke loose. We were being attacked.

ATTACKED BY FLAMETHROWERS

The squad in front of us was hit pretty hard. The enemy used flamethrowers as some of the men dug in. I think we had two or three men killed from the flamethrowers. Command was going to move our squad up front to reinforce that squad the next night. Understandably, not everyone liked this idea, after the flame throwers had been used the night before. It was a relief to all of us when we were ordered to move out in the afternoon.

COMBAT FOOD RATIONS

During the days in combat, we ate what were called "Field Rations." These consisted primarily of "C" rations. They were cans about eight ounces each and were an assortment of meals such as meat and beans, beef stew, hash, spaghetti, meat and potatoes or eggs and potatoes. Most of the time we ate these cold or if we were lucky we could heat them with sterno. For the first thirty-five days in combat, I never remembered us having any hot chow from our company kitchen. As our time in combat went along, the army started to supply us with better rations. One of the better rations was called a ten in one. This supplied ten men with three meals for one day. They came in a large cardboard box and it was quite an improvement.

RAINING SCHRAPNEL FROM TREE BURST

As we started to push inland, we met with sporadic resistance. One day as we were taking a hill, we came under artillery fire. It so happened that some of us in our company were in an area covered with trees. This can be very serious as a lot of the shells would hit the trees and the shrapnel would go down. I can still remember lying on the ground with shells exploding in the trees. I was lying between two of my buddies when they were both hit, one in the back and the other in the arm. Amazingly, I did not get a scratch.

ENEMY SOLDIERS PASSING IN THE NIGHT

As we moved inland it became apparent that the enemy was making itself scarce in the daytime. After taking a hill, we came to expect that a counterattack would occur that night. When it happened, it was always around midnight. One time we were ordered late in the day to move to another location. We didn't have enough daylight to reach our destination. We were marching on a dirt road in rolling countryside terrain when we stopped for the night. Usually at night, you always dug a foxhole, but due to the darkness that was coming on, we just left the road and lay out in the fields alongside. At around ten or eleven that night we heard voices talking on the road. The voices were Japanese. They were marching in the same direction as we were. It seemed that there were around one hundred fifty to two hundred men. Fortunately, we were all awake by then and nothing happened. If it did, it could have been complete chaos. At that time, our company was down to about one hundred and fifty men. This was not the last time that we had the enemy walk near us. There was another night when a Japanese patrol of about ten men walked within fifteen or twenty feet of us. We were spread out too thin and we did not know if there were more coming, so we let them go on their way.

MORTAR AND ARTILLERY FIRE TAKE THEIR TOLL

We had been fighting for over two weeks and were ready to capture the town of Rosario. This was a very important objective. The town was not very large, probably only a few thousand people. What made it important was the main highway from Manila to the summer capital of Baguio ran through it. We were dug in for the night when around eleven o'clock we were being attacked with rifle fire and mortar shells. It was complete bedlam as mortar shells were falling everywhere. I heard an explosion coming from the left of my foxhole, but could not tell how close it was. When daylight came, I saw that a mortar shell had landed about three or four feet from my hole. It ripped open my canteen and other things I had placed there. The shrapnel must have just missed me. One of the shells did land in the foxhole to my right about twenty feet away. Two of my best buddies, Travis Reid from Texas and Jim Lamont from New Jersey, were killed by it. I have many times thought about that night and about how close that mortar shell came to landing in my foxhole.

I had one other scary experience. We were dug in on a hill that was not too steep. A hedge about five feet high was growing along with a fence that was falling apart. Most of our squad had dug their holes just behind the fence. It was just starting to get dusk when Japanese artillery opened up on us. We did not receive too many rounds, but one of their shells landed about twenty feet behind us. I was trying to keep my foxhole buddy calm during this barrage, but when this close one went off he just panicked and I had to call the medics who moved him out. I never did see him again.

It was around midnight when we started picking up rifle fire. The enemy knew where we were from the artillery fire we had received earlier. We were prepared for an attack and it came. I was looking through the hedge in front of my hole and all of a sudden, there was a bayonet in front of me. The hedge had kept the soldier from reaching me. I fired my gun and waited to see what was going to happen next. Things quieted down as the enemy retreated.

SUICIDE BEFORE SURRENDER

It could not have been more than about ten minutes when there was a loud explosion that came from the other side of the hedge. In the morning, we looked and saw several enemy dead in the area. The enemy soldier in front of my foxhole had pulled down his pants and put a hand grenade on his stomach and blown himself up. We found this to be typical when enemies were wounded and did not want to be captured. Up to this time, we still had not taken any prisoners. We had many other nights when we were attacked, but the previous encounters were the most harrowing experiences.

Another night, we were on a hill and had just dug our foxholes when there was a loud noise like a train; this was followed by a huge explosion that shook the hill. This was followed by a couple more and they seemed to be coming closer. We found out that the shelling was from a three hundred MM howitzer. The shell from the gun weighed twelve hundred pounds. We thought that the enemy might attack that night, but it started to rain. It rained so hard that night that our foxhole filled with water. According to my Division History book, this occurred on February 3. Around the thirteenth of February, we were relieved by the 33rd Infantry Division. We were then moved south to the Santa Barbara area. We were there for rest and relaxation and to receive replacements. At this time our Division, the 43rd Infantry Division, was short two hundred and fifteen officers and thirty-eight hundred enlisted men. We wound up with fifty-three officers and one thousand eight hundred and fifty enlisted men. We also had about six hundred previously wounded men return to duty. It was nice to have hot meals in place of "C" rations.

Our rest and relaxation soon came to a halt. We were getting orders to relieve the 40th Division. The 40th happened to be the California National Guard. We were moving south towards Manila and to Clark Field. Clark Field was the main field for the US Air Force when the enemy had invaded the Philippines. The Japanese had a force of 12,000 air, army and navy all now reorganized as infantry. They had abandoned the field and moved west and north into the Zambales Mountains. Captured documents indicated that the enemy had constructed many fortified positions. The terrain had sharp ridgelines with steep gorges and deep ravines. Due to the terrain, we called in air strikes using Napalm.

While taking this hill, one event happened in our company I will never forget. Our company's first Sergeant had been overseas for two and a half years and had recently received a furlough to go home for a couple of weeks. While home, he got married. He rejoined us around the first of March, while we were in combat. We had just taken this hill and it was late in the day. The Sergeant was walking around inspecting our positions when a sniper shot him and his body rolled down the side of the hill. We retrieved his body and saw that he had been hit in the head and died instantly. Everyone was really shook up, as he was well liked. I still remember him to this day. We had taken a few prisoners, our first ones, who were in horrible condition. We captured a supply depot that had cases of San Miguel Beer that we put to good use.

We were in the area for about ten days when we received orders to move out. We were going to Taytay. It is about ten miles east of Manila. The enemy defenders had evacuated Manila and gone into the mountains east of Manila. We heard that this consisted of about forty thousand men. Once again, we were faced with difficult terrain. We were going from one hill to another trying to locate and destroy the enemy. We were still having counter attacks at nighttime, but they were not as often and severe as they had been when we first landed. We were given three-day passes for a rest and relaxation camp set up in the rear. I don't remember exactly where it was located. We had fresh, hot meals while there. I was there on April 12th when we heard that President Roosevelt had died.

CASUALTIES MOUNT AND HOSPITALIZATION

When I returned to my company, I noticed that another man and I were the only ones left out of our squad of twelve men. The company had been reduced to about fifty men from a normal compliment of about one hundred sixty. About a week or so later I was having physical problems. I became constipated very seriously and I was having eye problems. With this problem, I was sent to a field hospital, but they could not help my eyes. From there I was sent to a large hospital in Manila for examination. The hospital was next to Rizal Stadium and I was able to see a couple of baseball games. The hospital informed me that I had to be evacuated. So I was put aboard a hospital ship that was docked in Manila and sent to the island of Biak. It is an island off the coast of the western part of New Guinea. Being on a hospital ship during wartime is quite an experience as the ship is lit up like a Christmas tree at nighttime.

BATTLE FATIGUE AND REASSIGNMENT

I spent about two or three weeks in the hospital and had exams and psychiatric tests and sessions. It was determined that I had "Battle Fatigue." My case was determined to be severe, to the extent that I was to be reassigned to a non-combat unit. It was good news and bad news. It meant my combat days were over, but I had a sense of guilt that I was abandoning my "foxhole buddies."

I left Biak around the end of May and flew in a C46 to the island of Pililla for an overnight stop. The next day we flew to the island of Leyte in the Philippines. I was sent to a reclassification camp and after they reviewed my background, I joined the 1562nd Engineer Depot Company located in the town of Tacloban on the island of Leyte. We were located a couple miles outside of the city. The Fifteen-Sixty-Second and I were responsible for parts replacement for all of the mechanical equipment in the Philippines. We had a warehouse that was about one quarter of a mile long. I was assigned to the motor pool as a motor vehicle mechanic. It was a pretty good job compared to what I had been through. I was there about two months and got promoted to a T5, which was the same as a corporal. I was then put in charge of two other men. I was also assigned to guard about a dozen Japanese prisoners when they were used as laborers. They were captured when we invaded Leyte in October 1944. They were no trouble and probably very glad to survive.

Once I was able to borrow one of our large trucks and with about ten or twelve men in the back I drove the mountains to the west coast of Leyte to visit the city of Ormoc. The Filipinos were so happy to see Americans.

We slept in six-man tents that had wooden floors. We had mosquito nets and they were well appreciated. We also had a table

THE MOST BASIC OF WARRIORS, INFANTRY | 43

that was nice to play pinochle on. And we had electricity. When August sixth came around and we heard the atomic bomb had been dropped, we felt the war might end any day. There were other infantrymen in our company. One of the men in our tent was Emil Romer from Texas who had been with the 7TH Division that had landed on the island of Attu in the Aleutian Islands.

On August the ninth, the US dropped another bomb on Nagasaki. Finally, on the fourteenth, the Japanese agreed to surrender. We were so happy to have the war end that we stayed up all night playing pinochle.

WAR MATERIALS DUMPED INTO THE OCEAN

I can still remember going to the Tacloban and looking at the cargo ships in the Harbor loaded down with material and equipment dumping this cargo over the sides into the harbor. It was not needed anymore and not worth shipping back to the states. What a waste!

GOING HOME

With the war over the question was: when can we go home? A point system was put into place. It included length of service, months spent overseas and campaigns. My total was sixty-three points. With this many I figured to be on the way home by the end of 1945. Around the first of December I left the Fifteen-Sixty-Second and went to a camp set up in Tacloban.

On the twelfth of December, I boarded a Liberty ship that had been converted to a troop ship. Bunks made out of two by fours had been constructed in the holds of the ship. There were about eight hundred of us on board composed of about seven hundred Negro troops and one hundred white troops. We set sail for the good old

USA. We were fed only two times a day, which was fine as long as we were not doing any work. At the end of the day, the Captain would tell us how many miles we had traveled. We were traveling about two hundred and fifty miles per day.

The Liberty ship is built to carry a lot of cargo, not people, so due to the fact we were not loaded, the ship's propeller was not in the water all the time. This caused the bushings in the drive shaft to wear and caused the ship to shudder. We were approaching the Hawaiian Islands when the Captain decided to stop and repair the ship. We were floating in the ocean for two days while the crew replaced or repaired the bearings. Then it was off again to San Francisco. It was a wonderful sight and a great feeling to go under the Golden Gate Bridge.

We docked on January 7, 1946, my brother Jim's birthday. It took us twenty-eight days to cover almost seven thousand miles and if I recall, there was not one incidence of a fight on board the ship. We docked at the same pier that I had left on, but this time the pier had a huge "Welcome Home" sign on it. Our ferry took us up the Sacramento River to Camp Stoneman, the same camp I had left from over two years before. The first day there, the cooks fed us steak, the first steak I'd eaten in one and a-half years. I can't remember if I enjoyed it or not. From here, everyone was sent off to a separation center. I wound up at Camp Beale, which is near Marysville. Camp Beale is still used, although is now renamed Beale Air Force Base.

After getting to Beale, I called my mother and she was able to contact my dad, who drove up to Beale to pick me up. So finally, on January 13, 1946, I was a free man. My dad showed up with Chickie and Fred Carothers and my fiancée Artie Hornbacker in the car. I had not expected her to come, but it was wonderful to see her. I think when all servicemen are discharged, it's a very emotional transition.

A couple of weeks after coming home I was hit with Malaria. I never had an attack overseas, but when the war ended, it meant we were not required to take Atabrine anymore. Medication took care of the Malaria and I had one more attack after, but none since then.

WAR CHANGED ME, I WAS NOT THE SAME MAN

I didn't realize how much I'd changed until Artie and I decided to break up. I felt devastated, but it was the right thing to do.

I later found out some information regarding my old division that I had to leave due to the Battle Fatigue diagnosis. Their combat had terminated near the end of June. They were ordered to go to a winter camp near a town called Cabanatuan City. This is the area where the Death March from Bataan ended. In February 1945, Rangers had rescued the prisoners there. The 43rd Division was scheduled along with seven others to lead the invasion of Japan around the first of November. Instead, they were the first group to occupy Japan after the war. By the end of October, they were in process of being shipped back to the United States.

I stayed with my Mom and stepfather for a couple of months. The first few months it was very difficult to adjust to civilian life. I even went to San Francisco with a close friend of mine, Bernie Murphy. We looked into joining the Merchant Marines. I backed out, but Bernie, who had been in the Navy, decided to go to sea again. I bought a 1940 Nash car and got a job at Prolo Chevrolet in Santa Cruz working in the body shop. I finally felt that I had really returned to civilian life.

*Pete Conforti in 2014 speaking to the Ojai Valley
Veterans Memorial Day*

4

COFFEE GOES TO WAR, A MATTER OF LIFE AND DEATH

■ ■ ■ ■ ■ ■ ■ ■ ■ ■ ■

The most preferred drink of the military is not an alcoholic beverage, but coffee. In my family, no one ever drank coffee except my father, a veteran of the First Infantry Division. As a child, coffee was a foul-tasting brew offensive to the palate. Coffee is an acquired taste. Were it not for military service, I probably would never have acquired the habit. But this habit was acquired through very different circumstances than the mess hall meals.

My exposure to coffee and the habit that followed began on the slopes of Mt. Fujiyama in Japan, but not because of its taste, as was the case with confirmed coffee drinkers, or its popularity among older soldiers.

ARMY TRAINING CAN BE IRRATIONAL

Military orders and training don't always seem to make sense. While training in the hot and humid summer of 1951, military planners developed the idea of "water discipline." It was a concept that was supposed to teach soldiers how to control their thirst by limiting their intake of water when water was unavailable or in short supply in the field. Each morning, a water trailer would pull up into the designated mess area. Each soldier was allowed a half canteen of water that was supposed to last all day until the evening.

Infantrymen are human pack mules. They carry everything on their backs, hips or shoulders. That includes rifles, machine guns, tripods, rocket launchers and other necessities for the field. In the heat and humidity of Japan in the summer, they sweated profusely. We were always thirsty, as we were allowed only a sip of water at a time because we tried to make our water last through the day. Obviously, we were dehydrated as we sweated out more moisture than we consumed in water.

Most soldiers are loyal to their comrades and will strive to protect them under normal circumstances. As thirst developed during the day and many canteens were drained dry, a canteen left unattended during rest periods from the long marches was subject to theft. Soldiers were caught stealing water from buddies who still had some left.

ALLOWED COFFEE BUT NO WATER

At sunset and the end of the day's maneuvers, the troops rendezvoused with the mess trucks and field kitchens. Water discipline remained in force—and the men were very thirsty. The irony of the military planning was that we could have all the coffee we wanted.

Yes! We could have coffee but not water. Needless to say, that was the beginning of the breakdown to my natural aversion to that "foul brew" as I gulped down canteen cup after canteen cup of coffee. Coffee was wet and I needed something to slake my thirst, but coffee does not slake a thirst as much as a glass of ice cold water.

Returning to our camp, I gave up coffee. Coffee and I were destined not to be friends.

A DIFFERENT USE FOR COFFEE

My final surrender to the coffee habit, coffee's coup de grace, came in the trenches high in the TaeBek Mountains of North Korea, that mountainous spine that ran from Manchuria to South Korea. We were face to face with the Chinese Red Army. War is always dangerous, but more so for the frontline Infantry troops. Weariness from constantly being on the alert day and night took its toll. During the night watches it was almost impossible to stay awake and alert as enemy patrols probed out lines for weak spots and undefended foxholes and trenches. We were well aware of bunkers and trenches where soldiers who fell asleep were bayoneted or knifed. Even the worst fears can't prevent one from succumbing to exhaustion. Most soldiers at one time or another did fall asleep despite all the fears, threats, and official warnings. The body has physical limitations. It was only by luck and the Grace of God that there were not more deaths from infiltrating Chinese soldiers.

But slumber, no matter how desperately needed, could jeopardize the whole front line that you were ordered to defend. It could imperil not only your life and safety but the life and safety of all of those around you on the front line if the enemy punched a hole through your sector.

It was in those circumstances that I became a confirmed

"aficionado" of the "foul brew." It was then and there that my craving for coffee began. Heretofore, coffee was used to quench thirst or warm my hands and insides in frigid weather. The new approach to coffee related to its effects as a stimulant, and that's what I needed in the trenches while on guard duty in the wee hours of the night. Taste had nothing to do with this acquired habit.

POWDERED COFFEE, OUR SECRET WEAPON

Each box of C-rations came with canned food, biscuits, canned jam, candy, matches, sugar, salt, toilet paper, G. I. can openers, a can of fruit, and powdered chocolate. Not all C-ration boxes contained identical items, but all had packets of soluble coffee. Each night we collected several packets of dried soluble coffee for our guard duty assignment. As we stood guard in the foxholes and trenches in the wee hours before dawn and the natural forces of sleep deprivation began to overtake us, we opened our packets of soluble coffee and licked the raw coffee constantly until we were stimulated and wide awake. The coffee masked our bodies' craving for sleep. I believe that coffee saved many lives of soldiers who embraced this strategy for staying awake. But—they don't teach about this in any military manual!

With the combined experiences of using coffee to warm my hands in the bitter cold, and to quench unabated thirst, and to stimulate alertness on the battlefield, I became a confirmed and lifelong coffee drinker. It's a substance that helped ensure that I would have a long life.

SECOND COMBAT INFANTRYMAN'S BADGE

The Story of Bill Gates, Infantryman

Many men and women make the military service a career; however, as military service becomes more and more technical, very few military service personnel have ever been exposed to enemy fire. A prime example is an aircraft carrier with *thousands of support personnel and only a relatively small contingent of combat pilots. Most service personnel and technicians remain far from the battlefield and out of harm's way. However, that does not negate the fact that anyone who serves may be called to give his or her life, limb, and freedom for the greater good of the nation.*

Infantry, whether Army or Marines, remains the cutting edge of combat, along with fighter/bomber pilots. Armor and artillery is a close second, especially in the fluid situations of the battlefield. Rear area personnel may be caught up in combat when infantry, armor, and artillery are overrun by superior enemy forces, which occurs in all wars.

In wars past, Infantrymen were not accorded the status they deserved. Individual infantrymen were the least recognized branch of service. Often, the infantryman was a grunt, a dogface, or a ground pounder. Early in World War II it was becoming increasingly difficult to enlist men into the Infantry branch since of all military services, the Infantry took a disproportionate number of casualties, both the dead and the wounded. In addition, Infantrymen tended to suffer the most from the sheer physical strain of battlefield conditions in hostile terrain and extreme climate conditions. General McNair proposed that a special badge be created in recognition of Infantrymen who bore the brunt of battle and disproportionate mental and physical suffering. In 1943, the Combat Infantry Badge was created and later stars were added for each additional award for subsequent wars and combat service. But few men survived to enjoy that dubious honor.

Despite the recognition evidenced by the proud holders of the Combat Infantry Badge, silly Pentagon officials continue to refer to Infantrymen as "boots" or "assets" thus diminishing the status of "soldiers" as living breathing human beings with personalities, hopes, and dreams. This attitude has crept into the general population today where the "all volunteer army" has enabled the ordinary citizen to escape the worry and fear of military service. It has become the volunteer's problem.

It is little wonder that Combat Infantry soldiers do not rush to reenlist or make the military a career. Most Combat Infantrymen serve one or two hitches and return to civilian life with few honors and little recognition.

RURAL ROOTS

Much of the military manpower for World War II came from the small towns, farms, and ranches across America. The nation was still a rural economy. This had its advantages, since most farm boys understood outdoor living, the use of weapons, and navigating in the country. Bill Gates was no exception, spending his early youth on a sheep and cattle ranch in Colorado owned by his uncle. Ranching is a hard and rugged life. In addition to ranching, Bill's father was also employed by the Colorado Fish and Game Department. The harsh Colorado climate caused Bill's father to suffer chronic health problems and bouts with pneumonia. This necessitated a move to a more salubrious climate in Southern California in 1941. World War II was being fought in Asia and Europe at that time, but it hadn't affected the United States until that fateful day on December 7, 1941, with the bombing of Pearl Harbor. President Roosevelt and national planners knew that war was coming and the draft, war plants, and military preparation was evident throughout the nation. The Gates family lived in Culver City, with Bill attending Hamilton High School while his dad worked at North American Aviation Company as a machinist, just prior to that fateful day.

The story of William " Bill" Gates is the story of a Combat Infantryman who survived two wars, first in World War II, and later in Korea.

I received my draft notice while still in high school. Patriotism was sweeping the nation and I was no exception. I enrolled in ROTC and remained in it for two and a half years prior to being drafted.

In 1943, all high school students subject to the draft were called to meet in our library. Students who had been notified to report for induction were dismissed. I was the only one in that group assigned to report for induction, much to the surprise of my classmates. I left the fun and games of my senior year to the boys who remained. I wanted to go and soon was on my way to Fort Brady, out near where March Airfield is today.

Fort Brady had been a Civilian Conservation Corps Camp during the Great Depression and the new recruits opened the camp for Basic Training. Before long, I was traveling to the East by train to Northern Michigan to join the 33rd Infantry Division, a newly activated National Guard Division from Chicago. Draftees from California, Oregon, and Washington soon fleshed out the Division. Training continued in three-foot-deep snow. Some California men had never seen snow, and were unfamiliar with the intense cold of Northern Michigan. My experience on the sheep and cattle ranch in Colorado prepared me for the cold-weather training.

In addition to continued Infantry training, I helped guard the Sault Ste. Marie Locks, manning the torpedo nets that had to be pulled shut periodically to protect the locks from sabotage on both the American and Canadian sides of the border.

While completing a three-day Infantry problem, we were ordered to Camp Kilmer, New Jersey, where we received crew-served weapons. The unit was re-designated from the 131st Infantry Regiment to the 156th Infantry Regiment. I was assigned to Company M, a heavy weapons company that included 81 mm mortars and heavy, water-cooled machine guns.

BACK IN CALIFORNIA AND THE SOUTH PACIFIC

Patriotism runs in families. The Gates family, my family, was no exception. My brother enlisted in the Infantry and was sent to the South Pacific with the 1st Cavalry Division. He served with a heavy mortar company as a forward observer directing mortar fire. He fought in the Marshall Islands and Luzon in the Philippines. He received a severe head wound even while directing fire and continued to fulfill his mission despite his wounds. He received a Silver Star Citation for gallantry in action. Upon recovering, he was sent back into combat where he was wounded a second time and awarded his second Purple Heart.

ALONE WITH GERMAN U-BOATS

Boarding ship in New York Harbor, we set sail in a large naval and troopship convoy heading northeast towards Nova Scotia when the steering and propeller mechanism of our troop ship failed and the ship was detached from the protection of the convoy to make its way into Halifax Harbor. Considering the German submarine menace, a lone troop ship was a tempting target in the early days of the war.

The original destination was to be North Africa; all that changed and instead, we boarded the Queen Mary Ocean liner, a fast troop ship converted from a luxury liner. Once again the unit reboarded with Canadian troops for the voyage across the submarine-infested waters of the North Atlantic. We arrived safely in Glasgow, Scotland. From there we took a train at night to Devon, England where we were greeted by friendly Brits who appreciated our being there.

DEADLY TRAINING, THE SLAPTON DISASTER

I had witnessed a closely guarded military secret of World War II. I saw LST ships entering the harbor at Devon with extensive damage including the front bows shot off in what history has revealed as one of the greatest training disasters of World War II.

Off the coast, naval units were practicing for the Normandy Landings on D-Day. Extensive allied radio communications alerted German torpedo boats operating off the coast of Devon in Lyme Bay that something big was happening. Soon they found eight LSTs and one other ship preparing for amphibious maneuvers—tempting targets for the German torpedo boats! They struck on April 28, 1944 several days before the D-Day landing in June. The result was an unmitigated disaster, with 749 American soldiers and sailors killed and many more wounded. The result of the German speedboat attacks was the worst training disaster of the war, but it was the result of live enemy fire, not training. I saw the damaged LSTs as they limped into the harbor the next morning. The military censored every aspect of the disaster and threatened any soldier who dared to talk of what happened with court martial. Meanwhile, bloated bodies of American Infantrymen washed ashore. Dark rumors have continued to circulate that the disaster was really from friendly mistaken fire and not enemy action.

Meanwhile the Infantry marches, amphibious maneuvers, and training continued up to deployment to the Omaha Beaches in France. We learned how to prepare satchel charges for use against enemy-fortified bunkers as well as how to use Bangalore torpedoes against barbed wire entanglements. One of our best sergeants was of Polish background and was demonstrating the use of the satchel charges when the firing device caught on his jacket, exploding before he could extricate himself. He was a good man and highly qualified sergeant. The troops learned to attack straight up cliffs in anticipation of the soon expected assault on the German-held beaches of France.

MORE FRIENDLY FIRE AND INSTANT DEATH

While training with two Infantry companies, Dog and Easy Companies, the troops were engaged in a maneuver called "Pop and Crack," involving live overhead fire. Dog Company was below the hill and Easy Company was on the hill. Dog Company was using 30 and 50 caliber machine guns, Browning Automatic Rifles, and M-1 rifles. A confused lieutenant suddenly ordered all the weapons to open fire on the hill occupied by Easy Company, the very hill I was on. He apparently didn't realize that friendly troops were still on the hill. Suddenly, bullets were crashing around us in a maelstrom of copper and lead. Men threw themselves on the ground. The platoon sergeants were killed and troops were being shot to pieces. One man had a bullet through his teeth; a man next to me had his knee shot through. By the time the lieutenant realized his mistake, 15 or more men lay dead and many more wounded. A simple error that occurred for only a few minutes, but it was an unmitigated disaster for us.

That night I stood guard over a garage improvised as a temporary shelter for the American dead. No one else wanted the job, so I volunteered.

SHAEF HEADQUARTERS (1943-1945)

For some reason, the mission and unit designation was changed. We were given the task of paroling the moors, brush covered areas where German pilots had parachuted when shot down. Other assignments included guarding the docks, and eventually, we were assigned to guard SHAEF Headquarters, which was the Supreme Headquarters of the Allied Expeditionary Force in Bushy Park in England. There I had an opportunity to see all the major commanders of the American forces including Generals Eisenhower, Patton, Omar Bradley, Montgomery, and Biddle Smith.

From there we sailed to France several days after the Normandy landings when the beaches were secure. Attached to SHAEF Headquarters, we were involved in constant guard duty where I was assigned as a jeep driver for a Lt. Colonel. We set up in Versailles, France. I drove and he rode "shotgun," occasionally returning sniper fire. On one trip to the Battle of the Bulge area to a replacement camp, we saw soldiers in white uniforms on the road camouflaged against the snow. Suddenly, we realized we were lost and actually behind enemy lines as the soldiers in white opened fire on us. I made myself as small as possible behind the steering wheel as I raced out of harm's way with bullets whistling all around me. Later, we moved up to Brussels where we were engaged in quite a firefight, thus earning me my first Combat Infantryman's Badge.

A MILITARY ROMANCE

Not all service is pain and stress. While in Paris I met another corporal, not an American but a British woman soldier. She served with British G-3, plans and training. We dated for a while and married September 28, 1945. She immigrated to the United States as a war bride, but a war bride who had been a British soldier.

I'M IN THE ARTILLERY NOW

Upon entering German occupation after the war, I was assigned to locate and identify unexpended mortar and artillery rounds left over from combat operations. We would go through the fields and when we found rounds of ammunition, we would stake them for disposal by specially trained ordnance personnel. It was there that I supervised German prisoners of war and became acquainted with a German sergeant who had fought us in Europe. He was really a farmer who had to join the military. We had free-ranging discussions about what had happened in the war and got along quite well. However, he was not particularly fond of the youthful prisoners of war because of their hardcore Nazi orientation.

PEACETIME IN THE STATES

When I returned to the states in 1946, I reunited with my parents after a three-year absence. They hadn't seen me since I'd left high school. Now I was returning with a new British War Bride and ready to enter the world of work and industry. As a driver in the military, I had some peacetime employable skills and soon found work as a driver hauling Chevrolets from the factory to dealerships in California, but I was drawn back to the military as a part time soldier in the California National Guard where I joined and eventually became a sergeant in M Company, which was the heavy weapons company of an infantry battalion. Once again, I was in the Infantry. Little did I know or anticipate that soon I would be serving in combat in a new war on the far-flung outer edges of Asia. There I would be an occupation soldier in Japan before sailing from Yokohama in the dead of winter for Inchon, Korea.

I was with the heavy mortars, a 4.2" power-packed weapon equivalent in explosive power to a 105 Howitzer Artillery piece. The 4.2 Mortar is a high-trajectory weapon. Actually, it's two millimeters larger than a 105mm artillery howitzer round. It was referred to as a "Four-Deuce." It had a rifled barrel rather than fins, to promote stability in flight. Over time and with the addition of propellants, the range was increased to two miles.

Motor transport was essential because the Four-Deuce was too heavy to be carried as an Infantry weapon, even though it was included in the Infantry Table of Organization. The whole weapon with barrel, base plate and legs weighed about 330 pounds, certainly too heavy for light Infantry.

In the course of my deployment, the enemy always seemed to have the dirt supply roads under observation, with vehicles being a preferred target. That included our weapons company vehicles as well as tanks and even mess trucks.

Toward the end of my tour in Korea within a few days of my rotation home, several of our vehicles were damaged due to enemy artillery/mortar fire. We needed those vehicles if they could be repaired but if not, we could cannibalize parts to repair other trucks. As we worked to tow the vehicles back to our base of operation, we were subjected to enemy shellfire. Suddenly, I felt like a mule had kicked me as small pieces of shrapnel hit me. I was injured in two places but able to continue my work recovering the disabled vehicles. I bandaged myself but when I returned to my company area, I did not report to the medical station. That may sound odd, but I was due for rotation and if I was held back because of the injury, it would delay going home. I really wanted to go home to my wife and children. Under the circumstances, the war had not been abated even though there were "Truce Talks" that had been going on for months.

Years later, the shrapnel in my wrist began to give me trouble. Reporting to the Veterans Administration Clinic, they x-rayed my wrist and confirmed that shrapnel was still embedded in the wrist. However, the army did not recognize that as a war injury since I did not report it. A friend tried to get a Purple Heart for me since the VA report clearly stated the nature of the injury, but the Army ignored the request even though it was medically documented.

THE GUARD REESTABLISHED IN CALIFORNIA

My experience in the California National Guard was much more intense and significant than my service in World War II in that there was a closer bond between the men since we had trained together for one and a half years before entering combat operations in Korea. We formed lifelong comradeships that endured even as the veterans of the 160th Regimental Combat Team formed the Korean War Veterans Association.

When the 40th Division was reactivated as a California National Guard unit, sergeants such as myself with years of military service and actual experience were encouraged to apply for direct commissions in the California National Guard. With service as an Infantryman in two wars, I was well qualified. But the stateside Guard is often rife with politics, very different from the Guard on active duty and in combat, where ability usually takes precedence over politics. During wartime, incompetent Guard NCOs and officers were quickly sacked if they could not perform their duties during the intensive training and combat assignments. When I applied for a commission, politics and personal friendships resulted in my promised promotion being given to a less qualified individual. With that I left the Guard, but not my contacts and associations with my comrades from Japan and Korea. I was always proud of my service in the U.S. Army Infantry, and devoted to my men.

It is said of combat that you are never as alive as when you face imminent death. The bonds of comradeship forged in the battle are bonds that endure throughout a lifetime.

My English War Bride and I had a long and happy marriage.

Together, we had three children. She passed away several years ago and I remarried Frances, a widow who became my staunchest supporter as I turned my attention to the 160th RCT and 40th Division reunions. Today, we have a blended extended family and we both honor our former spouses. We have a mature understanding of life and the circumstances we encounter on this journey.

    I occupied a key role in the development and maintenance of the 40th Infantry Division War Memorial at Camp Cooke, California, which later became the Vandenburg Air Force and Missile Base. I helped veterans with placing memorial bricks at the Monument with their names and units from Korea. Frances and I always participated in the reunions, where I got the chance to renew friendships with old comrades.

Bill Gates passed away several years ago, but many veterans he befriended and assisted remember him fondly. Bill is one of the few men who served in Infantry combat at the Battle of the Bulge and in North Korea, earning the double Combat Infantryman's Badge and managing to live through both to be able to tell his story.

6

INFANTRY EQUIPMENT IN WORLD WAR II AND KOREA

■ ■ ■ ■ ■ ■ ■ ■ ■ ■

Infantry equipment was essentially the same for World War II and the Korean War. As recorded in the story of Pete Conforti, the United States disposition of World War II equipment was to destroy, discard, or simply dump the war materials into the ocean after the war. With the development of nuclear weapons by the United States, it was believed that conventional warfare of massed armies and navies was obsolete, since the United States and Britain controlled the atomic weapons. In contrast, the communist powers continued to expand and develop their military presence throughout Eastern Europe, China, Korea, and some Latin American countries. Free Western nations were subject to communist propaganda and espionage whereby American traitors soon acquired atomic secrets and delivered them to the Soviets. In addition, communist ideology penetrated the thinking of many Americans.

When communist North Korea, a client state of the Soviet Union, invaded South Korea, the conventional military forces of the United States were woefully undermanned, underequipped, and under trained. The United States was caught in a most difficult situation as Truman ordered American troops into South Korea to halt communist aggression. The American troops fared badly, lacking adequate crew-served weapons and modern armament. Reserve and National Guard armories were stripped of crew-served weapons for the troops fighting in South Korea. American soldiers paid a dear price with their lives by the misjudgment of military planners and politicians. As American soldiers' lives were sacrificed, they bought time for the American defense establishment to begin the task of rearmament and retraining millions of Americans called up in the draft.

Despite the fact that America had disarmed after World War II, American-grown communists were trying to label America as the aggressor and militarist. During that time, I witnessed communist demonstrations against a National Guard parade in Los Angeles. They were denouncing the Guardsmen as militarist and warmongers as they followed the parade down the sidewalks.

Today, the hazards of combat are still with us, but the modern soldier or Marine is far better equipped with modern equipment than the soldiers of past wars. The big difference is that there is no draft. Only volunteers remain in harm's way. Wars and the plight of American combatants are of little concern to the average American unless he or she has a relative or loved one in the armed services. Politicians are in a safer political decision-making mode when they don't have to worry about the general population having to share the burden of "their wars." Another factor is that few politicians have ever served in the military and are at an experience disadvantage in understanding their own foreign policy decisions that may require combat.

The following list of infantry equipment for that World War II and

Korean era describes what most soldiers were issued, whether they were Infantry or not.

INFANTRY EQUIPMENT AND SMALL ARMS

The M-1 Garand Rifle. This rifle was the basic Infantry weapon for World War II and the Korean War. It fired an eight-round clip and was equipped with an eight-inch bayonet. Basic Infantry had a cartridge belt, which contained six clips, or 48 bullets. The rifle was semi-automatic meaning a round was fired with each squeeze of the trigger. The M-1 weighed about ten pounds.

The cartridge belt carried a first aid kit, which was just a large bandage. An entrenching tool or a machete could be hooked on the belt. An additional hook held a canteen for water.

The entrenching tool was the second-most important piece of the rifleman's equipment. The entrenching tool could be used in hand-to-hand fighting if necessary; however, its main function was to dig foxholes and trenches for defense. Every time an Infantryman came into a position, he started to dig a foxhole for protection. The entrenching tool was first used in Hebrew warfare in the days of Moses, who commanded that a spade be attached to the end of every spear. The secondary but ultimately very important use of the entrenching tool was to bury human waste. It was a primitive but effective tool for field sanitation. Just think how advanced Moses was when even in the 20th Century some armies did not practice field sanitation.

Canteens contained a canteen cup for water or coffee, etc. Being aluminum, it was suitable for water only because some liquids reacted with the aluminum strongly enough to make a soldier ill. The canteen and cup were contained in a canvas carrier that hooked to the cartridge belt or pistol belt.

Both cartridge and pistol belts were developed to carry knives, machetes, canteens, first aid pouches, etc. The difference was that the cartridge belt held M-1 clips whereas the pistol belt held a pouch for pistol magazines or carbine magazines.

Carbines M-1 and M-2. The carbines weighted about five pounds and could be used with a short bayonet. The M-1 Carbine used 15 round magazines and was semiautomatic. The M-2 carbine held a 30 round banana magazine and had a selector lever for either full automatic or semiautomatic fire. The bullets were smaller than the M-1 Garand Rifle and didn't have the stopping power at ranges above 100 yards. After that, the quilted uniforms of the enemy soldiers could absorb some of the bullet's impact. For close-in fighting, the M-1 Carbine was a very reliable weapon. One of the Veterans' Stories tells of a lieutenant in the 40th Division who assaulted a Communist position alone with a carbine and 360 rounds of ammunition. He received the Silver Star for that lone action.

Some soldiers didn't like the M-2 carbine because the magazines would fail due to dirt or ice that jammed the spring that fed the rounds. Soldiers who carried the M-1 or M-2 Carbine used a pistol belt.

The 1903 Springfield Rifle. Every rifle squad had one soldier armed with the 1903 Springfield rifle, which saw service in World War I. It was a bolt-action rifle that required each round to be inserted in the firing chamber after each shot. It was equipped with a scope and was used as a sniper rifle. It fired the same round as the M-1 Garand rifle but used a five shot clip bolt fed. Originally, the rifle used an 18-inch bayonet that could be unwieldy in tight quarters. By Korea, they all used the 12-inch bayonet.

45 Cal Pistol. This pistol was developed for service in the Philippine Wars for fighting the Muslims in Mindanao around the early 1900s. The impact of a bullet was sure to stop an enemy in

his tracks. The previous pistols and revolvers did not have "stopping power." The 45 was magazine fed and semiautomatic. Gunners of crew-served weapons such as machine guns and mortars usually carried a pistol for self defense at close quarters. Officers often carried a pistol.

Mess kit and related equipment. The mess kit is a two-part folding aluminum pan that is used when soldiers are being fed in field kitchens. The canteen cup is used for liquids. The mess kit contains an aluminum knife, fork, and spoon. In combat operations where field kitchens do not exist, the mess kit is discarded and the soldier retains the spoon which is the only item kept for C-Rations.

Food rations are categorized as A, B, and C rations. "A" rations were the rations served in permanent mess halls and bases where fresh meat, vegetables, and baked goods are served. "B" rations were used in field camps and bases with semi-permanent mess halls, especially overseas where powdered eggs, powdered potatoes, and other foods were reconstituted and heated for approximate regular meals. That included dried milk, coffee, etc.

Class "C" rations were served to troops when no regular mess facilities were available. They came in boxes that contained three cans of food such as hash, beans, spaghetti, etc. A round of powdered chocolate or jelly candy was usually included. Each box had a can of fruit. Toilet items included a small can opener, matches, toilet paper, salt, pepper, etc.

MRE Rations. Modern military service people are much more familiar with the MRE rations that can be reconstituted by water and heating to approximate the heavier C-rations of previous wars. The MREs can last a long time in storage in almost any weather condition. Many civilian outdoorsmen and women can find comparable lightweight food rations in modern sporting goods stores.

Dog Tags, Every soldier wears "dog tags" to identify them if they are killed or wounded. The "dog tag" has the soldier's name, serial number, blood type and religious affiliation. The religious affiliations were usually Catholic, Protestant, or Jewish.

Religious Testaments and Bibles were distributed to the troops during war periods. Two religions recognized included Christian or Jewish. The military establishment supported the maintenance of religious faith. Today, there are increasing restrictions on the Christian faith to conform to the modern politically correct multiculturalism.

Other Infantry weapons included bazookas, from the 2.36-inch bazooka to the newer 3.5-inch bazooka that was eventually standard in the Korean War. Machine guns included the 30-caliber air-cooled machine gun that could be mounted on a tripod or used as a shoulder weapon supported by bipods in the front. Weapons companies had the 30-caliber water-cooled machine guns. Mortars ranged from the 60 mm mortars in the basic rifle companies to 81 mm mortars in the weapons companies. Each regiment had a 4.2-inch mortar unit. Other weapons included the recoilless rifle which was shoulder fired but had a back blast that gave their position away the instant it was fired.

Though they weren't equipment, Military Chaplains were an integral part of each unit. They did not carry weapons but many died in World War II and Korea, including Chaplain Robert Crane, whose Army service is described in the first chapter of this book. Military Chaplains are not trained in exclusive-denominational ministerial functions. Any chaplain may provide spiritual comfort to any soldier in need of religious support. Remember the story of the four chaplains of different faiths who died on the U.S.S. Dorchester in World War II. Military chaplains are much more ecumenical than civilian clergymen.

7

A MEDIC IN CHINA, A WAR AGAINST DISEASE

■ ■ ■ ■ ■ ■ ■ ■ ■ ■

The Story of Jack Schultz, Army Air Corps India, Burma, and China

Jack Schultz served in the India, Burma, and China campaigns during World War II. These campaigns were important lynchpins in the fight against the Japanese war machine. Most of the history of Eastern Asia has been forgotten or neglected except for the few survivors of the campaigns in India, Burma, and China.

Historically the Japanese home islands were deficient in most of the raw materials necessary for developing a modern nation. Japan had been closed to Western nations for centuries until in 1854 when Commodore

Perry of the United States Navy was sent to Japan to force the Japanese to open trade. Japan was a backward country, but upon learning of the advanced technologies of the Western world, Japan developed quickly as a modern nation and turned her attention to expansion in Asia as a source of raw materials. Korea and China were in the Japanese sphere of influence and control, but this was contested by Russia. Japan defeated China in 1895 and later Russia in 1905. Korea and Manchuria were annexed to Japan. During the 1930's Japan waged a protracted war against China while developing decidedly anti-Western themes such as "Asia for the Asiatics" and the East Asia Co-prosperity Sphere. However, most of Asia was controlled by Western European colonial powers. England controlled India, Burma, Malaysia, Australia, and Hong Kong. Even Portugal had enclaves carved out of China in Macau. The Dutch controlled Indonesia and the French had French Indochina, now called Vietnam. Japan was bereft of colonial possessions and was shut out of access to needed raw material. The doctrine of imperial expansion brought Japan into direct conflict with Western powers including the United States, which controlled the Philippines and Guam. The United States fleet stood ready for any eventuality at Pearl Harbor in Hawaii. Secretly, the Japanese Imperialists planned to neutralize the threat of the American fleet. The history of the attack on Pearl Harbor is well known.

The Japanese assumed that the battleship was the core of naval power and when Pearl Harbor was attacked, the Japanese attack bore down on the American battleships. But Aircraft Carriers and submarines rewrote the history of naval warfare in World War II. Most often, battleships were relegated to offshore gun platforms for assaults on enemy-held beaches.

The Japanese military machine was quick to take advantage of their temporary success at Pearl Harbor and overran most of the Far East: defeating British forces in Singapore and Burma, capturing and destroying the American army at Bataan and Corregidor in the

Philippines, and neutralizing French and Dutch forces in Indochina and Indonesia. The Japanese war in China continued even in the midst of a civil war between communist Chinese and the Nationalist government of Chiang Kai Shek.

The only nation capable of assisting Nationalist China was the United States. Chinese coastal cities were under the control of Japanese forces. The only door open to support the Chinese war effort was the air route from India/Burma over the Himalayan Mountains. Jack Schultz was part of that China experience as an Army Air Corps medic.

JACK SCHULTZ'S STORY

My story begins in Detroit, Michigan where I was born in 1922 and later attended local schools. Detroit was the automotive industrial powerhouse for the nation and continued to be for many years until after the Korean War. Upon graduation I was employed in the mail department of Detroit Edison Company, operating an IBM card tabulating machine in the meter-reading department. This was old fashioned and primitive compared to modern technology, but at that time, it was on the cutting edge of office engineering.

In February of 1943, I received greetings from my draft board and was assigned to Camp Custer in Michigan. Initially I was in a holding pattern so common in military service, where every soldier is familiar with the phrase, "Hurry up and wait." Camp Custer was a training center for troops inducted from the Midwest and was designed for company-level Infantry training. I remained there until the army had inducted enough IBM machine tabulators to form a training company. While waiting, it was typical boredom interspersed with exciting events. It seems that the army didn't know how to use an IBM computer operator in those primitive days when computing was a little-understood activity. Later, we were sent to Petersburg, Virginia for basic training. Many of the troops were drafted postal workers. From there we went 14 miles down the Potomac River to learn the army way of tabulating. We were informed that in the army, we no longer had any civil rights and were under military law. The six weeks at that station were miserable, especially in July with 90-degree heat and 90% humidity. As surplus personnel I was sent to Fort Mason on the San Francisco Bay for three months where my duty assignment was to file thousands of personnel tabulating cards, one for every soldier who left the West coast of the United States.

TRANSFER TO TORRANCE AND ITALIAN POW'S

After all the tabulating training to learn the Army's way, I was reclassified as a clerk typist. Along with a squad of soldiers I was then sent to a hospital near Los Angeles and assigned to the Receiving and Disposition office. We didn't have to do KP or other menial chores while stationed in Torrance because we had Italian prisoners of war to do that work for us.

Previously, the Italian POWs had worked in the scorching hot deserts of Arizona picking lettuce. They were happy to be assigned to Torrance where the working conditions were much better. Though they were POWs, many of them had relatives in the Los Angeles area and on weekends, their relatives would drive to Torrance and take them home for the weekend. No armed guards, barbed wire, or machine guns! I thought, "What a way to run a prison camp!"

INTERESTING "OLD ARMY" CHARACTER

I met some interesting characters from the "old Army," including a former Master Sergeant who ran afoul of the system while supervising an Army movie theater. It seems that the sergeant was caught defrauding Uncle Sam of ticket sales, resulting in the sergeant being demoted to Private after serving a tour in the Army's infamous Leavenworth Prison. But the sergeant was returned to active duty, a reformed Private with a past. Why waste an old sergeant? He was good for something. It was probably a better way of handling a corrupted man than the modern system.

A 5000+ MILE CRUISE IN ENEMY WATERS

The war was raging in Europe and Asia and I was still stateside with the rank of corporal. I wanted to advance in rank and saw no prospects for advancement, so I applied for duty elsewhere.

I was sent to Abilene, Texas where I learned to be a medic in six short weeks. After a brief furlough, I was sent back to Wilmington Harbor near Los Angeles for an assignment in the Far East. Five thousand troops boarded *U. S. S. Admiral Benson*, a relatively new troop transport designed to serve amphibious forces. It was in November of 1944 when we set sail alone and without any air or naval escort. We had to change course every seven minutes since the Pacific Ocean was patrolled by Japanese submarines even up to the coast of California. During World War II, the open oceans were a war zone. We changed course every seven minutes because that was how long it took an enemy submarine to lock onto us as a torpedo target. I was not a combatant, but the voyage from Wilmington, California to Bombay, India on the other side of the world was fraught with anxiety and concern about being torpedoed by Japanese submarines.

We stopped for a day in Melbourne, Australia. Melbourne is the southernmost major city in Australia, and faces the Tasmanian Straits and the Indian Ocean. Our ship took the long way around Australia since Indonesia and Southeast Asia were in the hands of the Japanese. We would have had great difficulty evading Japanese ships and planes in that area.

MORE TRAVEL BY RAIL, FERRY AND PLANES

Eventually, we landed in Bombay, India. Now called Mumbai, it was on the west coast of India, far from both our port of embarkation and our ultimate destination in China. I am sure that the whole planned route was for the safety of the ship and the thousands of troops on board.

From Bombay, we traveled across India to Calcutta. We had no contact with the Indian, and later the Chinese, populations. One of the major health problems in Asia was waterborne disease. As we traveled across India, the train would stop and we would offload to where huge iron kettles were filled with boiling water. We would fill our canteens from those kettles. It took a week to cross the midsection of India. Upon arriving in the outskirts of Calcutta, we boarded another train for travel to Assam. At the Brahmaputra River, we took a ferry to the other side before boarding yet another train that took us to Assam. Assam is an eastern province of India located just south of the eastern end of the Himalaya Mountains.

"SKYWAY TO HELL"

This was the point where American aircraft, mostly C-46s, were flying the "Hump," the G.I. pilot's name for the Himalayan Mountains, the highest mountains in the world and very treacherous for any aircraft. Up to that time, over seven hundred American planes had disappeared in that daunting terrain, many, if not most, were never found again. The pilots of that era referred to the Himalayan Mountains as the "Skyway to Hell."

Fortunately, by the time I took the flight from Assam to China, the British had driven the Japanese out of Burma and the more primitive C-47s, which were limited to16 thousand feet, had been reequipped with super chargers that enabled them to fly over most of the highest peaks. My flight was mostly over jungle areas. For those airmen who flew in the early days of the war it was a frightening experience flying over the Hump in turbulent mountain-spawned weather with unpredictable updrafts and down drafts and violent thunderstorms. Flying the Hump truly was a Skyway to Hell.

MY CHINA ASSIGNMENT

As a trained medic and clerk, I was being flown into Kunming, China as medic support for the American airmen and soldiers who were receiving the supplies destined for Generalissimo Chang Kai Shek's Chinese armies who were fighting the Japanese.

I was assigned to the hospital unit of the 14th Air Force, and was under the command of General Claire Chennault. My activities were not particularly exciting, but the travel through India, Burma, and China in that bygone era, was quite exciting enough for a young man from Michigan. It was so foreign and so different from any experience I'd ever had.

Once in China, we stayed close to base and seldom mixed with the local population except for one encounter while riding "shotgun" with a convoy. The roadside was filled with Chinese and while driving through that sea of humanity, a Chinese man attempted to climb into the back of my truck, not realizing we were there to protect our supplies. He must have had a rough landing when he was forcibly ejected.

DISEASE AS DEADLY AS BULLETS

Civilians who have never experienced the living conditions in Asia and many undeveloped countries at that time have no knowledge of the climatic conditions, the natural disasters, and the diseases that threaten human beings. Americans raised with a high level of disease control and prevention are particularly at risk in these backward areas. In so many wars, there have been more casualties to disease and weather than actual enemy bullets. Disease killed Alexander the Great at age 33. The famed German general Erwin Rommel was sick with a stomach disorder during the most critical battle in North Africa. Weather destroyed Napoleon's army in Russia. Disease killed more American soldiers in the Revolutionary War than British bullets and bayonets. Medical care and field sanitation are essential to maintaining the proficiency of any modern army in the field.

Most of our medical problems related to waterborne diseases and filthy living conditions that caused frequent intestinal problems among the troops. Our unit was involved in preventative medicine to maintain the operational status of the unit.

Being in a medical outfit may not be viewed in the same light as being in a combat unit. Yet, without medical care, the fighting strength of any unit can collapse. Treatment, shot records, and hospitalization records must be kept and evaluated. At that time, China had leprosy, elephantiasis, typhoid, malaria, and many other diseases. The medical corps of the United States Army Air Corps had a vital role in maintaining the health of our pilots, soldiers, and ground crews.

With the dropping of the atom bombs and Japan's subsequent surrender, our station was closed and we traveled to Shanghai and thence to Seattle and home, where I was discharged.

LIFE AFTER SERVICE

I had enjoyed my time of service in Southern California, so I bought a car and drove back to California, where I have lived ever since. I continued to be employed in data processing and worked for different agencies such as the California Franchise Tax Board, preparing payments to returning G.I.s. (The State of California gave each veteran $20.00 a week for six weeks for transition money until they could find employment.) I also worked for Lockheed Aircraft Company and many other firms. I attended the University of Southern California where I earned a degree in business and economics. I retired to Ojai, California where I married. As a veteran, I joined the veterans of Foreign Wars Post 11461 in Ojai where I support all veteran activities and causes.

FROM REFUGEE FAMILY TO PATRIOT AIRMAN

■ ■ ■ ■ ■ ■ ■ ■ ■ ■

The Story of Edward Kachadoorian, Army Air Corps

The life of Edward Kachadoorian began with his refugee parents entering the United States after World War I and struggling through the Great Depression followed by American industry gearing up for a total war effort. Before entering military service, Edward experienced the economic living conditions during both the Great Depression and the early part of World War II.

Throughout the world, refugees have sought the safety and freedoms of our American homeland. Originally, the Americas became dumping grounds for the unwanted religious dissidents, orphans, and poverty-stricken underclasses of Europe. This story is from the records and oral history of a family that fled the genocide of a very ancient Christian people of the Middle East. This is the story and experience of Edward Kachadoorian, the son of Christian refugees, and his service to their adopted country, the United States of America. No one dared speak ill of this nation in the presence of Edward, for he would challenge anyone who denigrated this country in his presence. With anger in his voice and tears in his eyes, he would recount how this nation was the only nation that provided a safe haven for the survivors of the massacres in the Middle East. He felt that he owed a debt to our nation. He helped repay that debt by enlisting in the United States Army Air Corps during World War II.

THE BACKGROUND OF THE GENOCIDE AND EXPULSION

The oral history of the Kachadoorian family details how an entire race and religion was exterminated in a country that prides itself on its liberalism and democracy, the Middle Eastern country of Turkey. How could a nation turn against its own citizens and create genocide against a minority race and religion that had inhabited Asia Minor for over 2000 years?

BELIEF IN THE SECOND AMENDMENT

Edward was a strong believer in the 2nd Amendment of the United States Constitution because disarming the civilian population of Turkey was the beginning of the genocide. In the name of national security, Turkey imposed arms control and demanded that the citizens turn in all hunting rifles, shotguns, pistols, and even daggers. Obedient to their government, the civilian population surrendered their weapons to government authorities. After all, World War I was being fought *and Turkey was allied with Germany in fighting the British and allied nations.*

GUN CONTROL, A PRELUDE TO GENOCIDE

Shortly after the Turkish government confiscated civilian weapons, Edward's family could hear gunfire down at the police barracks. The Turkish government had sanctioned the execution of the leadership of the Christian Armenian communities. Next, the government rounded up the men and boys between 16 and 60 and took them to the Army bases where, in groups of five or six, the Armenian men were told to run for their lives while the Turkish soldiers ran after them and used them for live bayonet practice. The "final solution" to this genocide was to have the women, children, and old people fall out into the streets with whatever they could carry. Once the Armenians were in the street, the Moslem population rushed into the Christian homes and took over their property.

The women and old people were marched far to the south without food and water. Each night, they were subjected to sexual abuse and murder by their captors until within a short time one and a half million people perished, including Edward's father's first wife and child. His father had escaped these massacres because he had immigrated to the United States before these atrocities occurred and was establishing a new home in New York. He had planned to send for his wife and child when he was financially able.

Edward's mother's family had a different story. They were educated people, unlike the general Muslim population. His uncle was a medical doctor and his immediate family was guaranteed protection if he served as a medical officer in the Turkish Army. However, his family was held captive on a Turkish Army base where they witnessed the killing of many Armenian men and boys. Upon secretly learning of the death of the doctor uncle, the small family fled to Constantinople, now called Istanbul, where they secured passage to the United States with little more than their lives.

DOCUMENTATION OF THE GENOCIDE

The Armenian Genocide was well documented by Henry Morgenthau, who wrote the history of that event with thorough documentation. He was the United States Ambassador to Turkey. Little did he realize that people of his Jewish background would undergo a similar genocide under Hitler years later! When the Nazis started their genocide against Jews, Hitler is quoted as saying, "Who remembers the Armenians?"

THE GREAT DEPRESSION

Immigrant and refugee families suffered more than most citizens during the Depression, since they were able to survive only by taking the most menial of jobs in a nation largely bereft of the social and economic safety nets now in place. There was resentment by Native Americans when foreigners took scarce jobs. There was abuse of the foreigners who were underpaid, were paid short wages, or were actually abused by employers. Edward's family survived by living as an extended family and sharing in the meager resources earned in the labor force. And, not all immigrant nationalities were friendly and cordial with one another.

WAR ACROSS THE SEA

The 1930s saw war in China, civil war in Spain, and war in East Africa. Those wars were preludes to World War II. Some American leaders saw the eventuality of the United States entering the war, but most Americans had been disillusioned with the aftermath of World

War I and the self-serving motives of the victors. There was a strong America First movement that wanted to avoid American involvement in the wars of Europe and Asia and the colonial policies of England, France and Russia.

FULL EMPLOYMENT, GOOD WAGES, AND SCARCE GOODS

As wars began far across the sea and a rising need occurred for America to prepare for national defense, America's industrial might began developing as never before seen in the history of the world. Jobs became plentiful for any able-bodied man and eventually any able-bodied woman, especially in industries heretofore closed to women. Wages were good. Overtime pay was available. Special bonuses were offered for Sunday and Holiday work, but with the bombing of Pearl Harbor, civilian needs for automobiles, housing appliances, etc. became secondary. Supplies of food, gasoline, and most manufactured items were suddenly rationed as supplies were diverted to the military forces and our allied nations. Everyone was issued a ration coupon book for meat, sugar, coffee, tires, appliances (if available at all), and canned goods. Gasoline was severely rationed except for those people in vital defense manufacturing jobs.

NOTHING WASTED

Today's conservationists are pikers when compared with the salvaging of everything imaginable during World War II. Wet garbage was segregated and trucked to the pig farms at the junction of the Los Angeles and Rio Hondo Rivers. Old newspapers and magazines were collected at the local schools to be remanufactured. Schools had contests rewarding the classes that brought in the most old newspapers. All metal, including tin cans, was recycled for war use. No metal was trashed; it was all recycled. Front lawns and empty lots were converted into "Victory Gardens." Even city folk were raising chickens and rabbits in their back yards to supplement the meager meat ration.

LOS ANGELES, AN ARMED CITY

The defense industries throughout Los Angeles were heavily defended with camouflage nets over entire factories. Barrage balloons floated above every major factory to prevent dive-bombers from attacking. Searchlights and antiaircraft guns were emplaced throughout the Los Angeles Basin. Nights were punctuated by air raid drills as volunteer air raid wardens toured their assigned city blocks to make sure all lights were off during blackouts. Where autos had been produced, the factories produced jeeps and tanks. Civilian production ceased.

EDWARD'S OWN STORY AS HE GOES TO WAR

I remained passionate about this country even as I was growing up in the Bronx as a streetwise New Yorker. In many ways, we, the immigrant children, had a street wisdom beyond our years. The Bronx was a mixed Irish-Jewish community. I learned the hardscrabble existence, earning money and working in areas now closed to modern youth because of child labor laws and OSHA regulations. I was fascinated with airplanes, especially Zeppelins. I watched the Hindenburg fly over New York City during the 1930s. Leaving high school, I enrolled in the Delahany Institute of Technology in Harlem. At the institute, I was trained in aircraft repair, and as I was caught up in the great epic struggle of World War II, I enlisted in the United States Air Force as a mechanic based on that training. Enlisting in Texas and training at Love Field, I had a relatively safe assignment compared to the combat-oriented arms, and the dangers of bomber and fighter crews, especially the 8th Air Force, which were taking horrendous losses in the air wars over Europe and Asia. But, nothing beyond our shores was safe during World War II. The oceans around our country teamed with enemy U-boats, particularly on the East Coast, where almost 400 American ships were sunk.

I was married at this time. Upon enlisting, the Air Force personnel learned of my musical skill, both voice and violin. They were going to assign me to the Air Force Band, but my wife was adamant that I be a "regular soldier," not a musician. However, my mechanic skills dictated that I should serve in that capacity rather than part of an aircrew.

Technically, I was still a child, like millions of other servicemen. Servicemen under 21 years of age could not vote, marry without parent's consent, or even sign a legal contract. Yet, most ironic, they could be committed to mortal combat, fighting, leading, and dying for the national preservation.

Ed and his wife in Texas

Eventually, my unit was sent by convoy and air to Sicily, Italy, and Morocco. Morocco was where the famous Casablanca Conference between President Roosevelt and Winston Churchill had taken place the year before.

THE CASA BLANCA CONFERENCE

That conference made policy regarding the prosecution of World War II. It was decided that the Axis forces would be driven out of North Africa and that this would be followed by an invasion of Sicily and later Italy. It was decided that nothing but unconditional surrender would be accepted from Germany. The Soviets were involved in this conference, and the policies were formulated.

While I was stationed there, Russian planes and crews landed and I took pictures of them. I didn't know what purpose the Russians had in Morocco at that time.

B17 Bomber in Morocco.

"LOOSE LIPS SINK SHIPS"

All mail and all communications from servicemen was carefully censored. The saying "Loose lips sink ships" was not a cheap slogan. Loose lips did sink our ships by the hundreds, even within sight of our continental shores. Crossing the Atlantic and Mediterranean Seas was fraught with untold danger and potential sudden death from German U-Boat attacks. Secretly, and on a very small pocket notebook, I kept a brief handwritten log of my crossing the Atlantic and subsequent destinations.

    I spent twenty-six days at sea sailing from Virginia across the Atlantic, through the Straits of Gibraltar, passing Malta and Sicily, and landing in Italy. I spent two days in Italy at the airbase northwest of Naples before flying to Algiers. Later, I spent two days in Algiers before flying aboard C-47s to Casablanca, Morocco, which was my final destination. My main duty station was Cazes Air Base for ten months, where I worked in a mechanical field.

While in Morocco, I had an Army Air Force buddy who was Moroccan in background. While most servicemen found parts of Casablanca unsafe, I was free to go almost anywhere as long as my Moroccan buddy was with me. I even ventured into the inner city and saw Moroccan life as few tourists or military people have ever seen it. My military service was routine but for a nineteen-year-old, it was exciting to see the activities of the military air bases and explore Morocco.

On the streets of Casablanca

Two events occurred in my life that were distressful. My father died that year I was at Cazes Air Base. I wanted to have leave to go to his funeral, but with the war raging in Europe and the Pacific, the Air Force could not let me go home. That was a sorrow that I never could resolve. The second event was the birth of my son. Once again, I was deployed and unable to go home until the end of the war.

My wife had a humorous situation as my son began to talk. But it was embarrassing to her. She had a picture of me in my uniform on the mantle. She would tell my infant son that that was Daddy. In those days, the millions of soldiers stateside were always in uniform. Apparently, the infant thought that any soldier in a uniform was "Daddy." When Louise, my wife, would be out in public, my infant son would see a soldier in uniform and began pointing and saying, "Daddy, Daddy." Some soldiers saw the humor. Other soldiers would react angrily saying, "I am not your daddy!" This was especially embarrassing if the soldier was with a wife or girlfriend and it was doubly embarrassing to Louise, my wife.

Bomber at Caze Air Base in Morocco.

With the end of the war in Europe we were transferred to Asia, since the war with Japan was still raging. We flew to Karachi, which was in India at that time but is now the capital of Pakistan, a Moslem country formed in 1947. After a flight to Calcutta in India I was sent to board a troop transport to sail to Hawaii and back to the United States.

The following is my simple hidden log I kept on my wartime travels. It is written as I recorded with italics for explanations.

MY SECRET LOG

Hidden on my person and secretly recorded is the log of my Atlantic crossing and subsequent travel:

Arrived O.R.D Greensboro, North Carolina August 15,1944. Left 1700 September 24th. Arrived at Camp Patrick Henry, Virginia September 25. Arrived 12:30 p.m. at dock. Boarded S.S. Felix Grundy. Set sail at 6:30 p.m. October 1, 1944

Everything ran smoothly on ship until 10-14-44 at which time at 5:00 p.m., a sub was detected on radar. At 8:00 p.m. alert sounded. Lights went out and everyone was ordered on deck with life jackets. Tankers and Liberty ship in flames. Tanker sank in 8 minutes. They were hit by torpedoes on October 15, 1944. Got word that the Liberty ship sank also. Subs were not sunk and –are still following the convoy.

October 16, 1944. Nothing happened during the night but subs still following us. We are expecting an aerial escort soon. Aircraft carrier has arrived and planes are patrolling the convoy. *(This traumatic experience haunted Edward for the rest of his life as he witnessed drowning troops in the open Atlantic and no ship in the convoy dared to stop to rescue them for fear they would be the next casualties of the German U-Boats.)*

October 17, 1944. Read book called *The Stuttering Bishop*. Very good. Nothing unusual happened. Carrier and planes left.

October 18, 1944 at 10:00 a.m. entered the Straits of Gibraltar. Land close by. It felt good to see land after seeing water for so long. The Rock of Gibraltar looks just like its picture. Took our money from us to change.

Sighted our destination which is Oran, North Africa and at 7:30 p.m. orders were changed to continue to Sicily—Do not know if we will stay there or go on. Most likely, we will stop for supplies and continue on.

10/20/1944. Have to get up for roll call at 7:00 a.m. Breakfast at 7:30 a.m. from now on. We were not allowed in the hole (*compartments*) until after supper because of a disagreement between the Air Corps and the A. F.C., which nearly caused a riot.

We passed Algiers between two and four p.m. Nothing unusual happened.

October 21, 1944 Had roll call at 7 a.m. and then P.T. (*Physical Training*) Breakfast at 7:30 a.m. This will go on every morning from now on.

Took our Atabrine tablets at 1 p.m. For malaria prevention, we will take a tablet every other day instead of once a day.

Got $4.00 of my money at 2:00 o'clock today—was only supposed to get $2.00 but they gave me $4 so that I could lend Morrey $2 because he is broke.

We have been seeing a lot of mountainous terrain all day along the coast of Africa. No cities at all could be seen.

Passed the beach of Bizerte.

Read *The Case of the Curious Bride* and *The Case of the Silly Girl* by Earle Stanley Garner and both very good. Nothing unusual happened today.

October 23, 1944. Lost sight of land sometime between twelve midnight and 7:00 a.m. this morning. Saw the Island of Malta and in the distance, Sicily.

October 24, 1944. Pulled into Port Augusta, Sicily. Saw some small fishing craft that looked like Gondolas with sails. Got there at 7:15 a.m. Left at 9:20 p.m. Looked like a modern city—saw Mt. Aetna in the distance. Was snowcapped. Learned that I am part of the mail-out for writing the most letters. Nothing unusual today.

October 25, 1944. We are sailing in the Ionian Sea and then got to the Adriatic Sea. Got to the border of Italy at 1:20 p.m.

October 26, 1944. Arrived at Bari. Got a haircut and shave—25 cents.

October 27, 1944. Left Bari by cattle car at 12 Noon.

October 28, 1944. Arrived at Coserta at 5:00 a.m. Went to town to get some souvenirs.

October 29, 1944. Went to town and drank some wine. Got sick on it.

October 30, 1944. Stayed on Post and saw picture called *Make Your Own Bed*. Informed at 5:45 p.m. was to ship out at 8:00 a.m. in the morning.

November 1, 1944. Left 12th Regt. Depot, Coserta, Italy at 8 a.m. Arrived at Cipocichint Airport at 9:45 a.m. Took C-47 #8577 at 3:15 p.m. Arrived at Algiers, North Africa at 9:20 p.m. Stayed overnight in staging area.

November 2, 1944. Left Algiers at 7:00 a.m. arriving at Casablanca at 11:45 a.m.

August 20, 1945. Arrived in Karachi at 10:30.

September 6, 1945 Left Karachi via C-47 in Agra. Left Agra arrived in Calcutta, India

I ended this log upon leaving India, but eventually, the troop ship took us back to the United States via Hawaii thus completely sailing or flying around the world in the course of my military service.

OBSERVATIONS IN INDIA

I had seen many primitive lands and living conditions, but nothing matched the poverty and squalor of India. India was still a British possession and as we flew into Calcutta, flying by leapfrog journeys across the Middle East and South Asia, I saw a world completely foreign and bizarre. I saw people so skinny they appeared on the verge of starvation. Yet, fat cattle roamed the streets and highways in large numbers, appearing well fed. I soon discovered that cattle are sacred animals in India and that they were not used for meat.

I saw the Ganges River with the burning Ghats where the deceased were cremated out in the open and then their ashes thrown into the Sacred River where the Indians believed the deceased will go to a paradise. But, some people did not have sufficient fuel to complete the process. The half burned bodies were thrown into the river. Reptiles supposedly feasted on the remains downstream. The city was filthy and totally unhygienic except for the special compounds reserved for military personnel and the British people.

Of course, the colonial rulers and upper-class people lived well amidst the abject poverty. India was still controlled by a caste system bizarre to the Western mind. The lowest caste were the untouchables, and even being in their shadow was cause for the upper classes to be alarmed.

THE POST WORLD WAR II YEARS

World War II ended the great Depression. Manpower and womanpower had been totally mobilized for war service and the defense industries throughout the country. Wages were good but there was a scarcity of goods, food supplies, and gasoline. Rationing controlled who got what each week and each month. People caught hoarding or going to the "black market" were severely punished.

THE COLLAPSE OF THE ECONOMY AFTER THE WAR

I'd been training in aircraft mechanics at the Delahaney School of Technology in Harlem. I spent my time in the Air Force as an aircraft mechanic. I'd been employed at Douglas Aircraft in Santa

Monica prior to enlistment. During the war, aircraft mechanics were in short supply and were needed in every aircraft company throughout the Los Angeles industrial centers.

It would seem that my skills and background with aircraft mechanics would ensure a good career upon my return from service. But, the end of the war caused a major collapse of the war manufacturing industries, especially in aircraft production. The mighty fleets of bombers, transport planes, and fighter planes were jettisoned into the ocean depths, mothballed out in the deserts near Tucson, Arizona, or sold or given away to foreign airlines or foreign governments. There was a giant pool of unemployed defense workers in the United States when I returned from overseas. Work was scarce and the threat of another major depression threatened. One saving factor was the G.I. Bill of Rights that created educational opportunities for returning veterans. Vast numbers of veterans took advantage of this education, which paid their tuition, books, and living expenses for them and their families.

In my case, I tried to find employment to support my wife and infant son. The work I found was low paying and temporary. I painted houses. I waited on tables. Like my immigrant family did when they first came to the United States, we survived by living together and sharing our meager resources. During the war, all my family had worked in the defense industry. Now they were all unemployed except my sister who worked for the Pacific Telephone Company.

One Christmas, Sears, Roebuck was hiring for the Christmas season. I applied and was hired in the toy department selling toys. Soon I was in the tool and hardware department where my mechanical skills were helpful in explaining mechanical things to customers. I learned that there were better opportunities for income and sales in the automotive department. I transferred and soon I became their leading salesman. I had found my niche. I was a salesman with critical knowledge of mechanics and tools that benefitted my customers.

Life was good and I finally joined the middle class, making a comfortable living. My wife worked as a waitress and was quite successful in her own right.

TRAGEDY UNRAVELS MY LIFE

Just prior to Christmas in the wee hours of the morning, I received a call from a hospital that my son had been in a terrible accident. He was driving a Volkswagen Beetle and waiting at a stop signal for the light to change when a full sized car hit him at 100 miles per hour from the rear. Essentially, my son was brain-dead, but I didn't know it at the time. I rushed to the hospital where he was on life support, with every tube and instrument of medical science hooked up to him. I stayed there in the lobby day and night hoping and praying for a miracle. I prayed in English and in my original language. After almost a week, even the machines of modern medicine could not sustain body functions and my only son, my child, died at the age of twenty. My wife and I were devastated. We were like lost souls. Our grief was inconsolable.

WE QUIT LIVING

I left my job at Sears, which had been one of the best companies in America and provided opportunities rarely matched by any other company. I pulled out my profit sharing until I could decide what to do. We found a business opportunity outside the Los Angeles area and my wife and I became comfortably successful. We consoled ourselves by helping our niece and nephews. They became our surrogate family, since we were now beyond the childbearing age. The sponsor company of our business saw our success and offered us employment as business brokers. There we became more than successful, since employment involved an understanding of mechanics and marketing.

CONCLUSION

Edward did not live to see retirement. He died early. Those who knew him knew that he died of a broken heart for the son he loved but could not replace. Despite his untimely death, he left a legacy of service to his beloved country, to his sister's family and children, and he had the admiration of customers who could always depend upon him to give the best of service beyond the simple act of selling. They trusted him.

Edward Kachadoorian separation paper.

9
AMPHIBIOUS OPERATIONS – WW II AND KOREA

In the wars of the second half of the 20th Century, many Army and Marine bases had large tower-like structures that seemed out of place in the open fields. In time, soldiers and Marines learned that these mockup towers were for the purpose of training troopers to climb off troop ships on nets suspended like ladders, and make their way down into landing craft that could carry a platoon at a time to amphibious assaults on fortified beaches.

The general public has the impression that amphibious operations are unique to the Marine Corps, but that is not correct, even though the Marine Corps perfected the use of amphibious landings. World War II in Europe saw many U. S. Army amphibious operations in Sicily, North Africa, Italy, and the beaches of Normandy. In the Pacific, the war in New Guinea and the Philippines was primarily carried out via Army operations. Joint Army and Marine landings occurred on Okinawa and later at Inchon, Korea. The landings at Tarawa, Guadalcanal, and Iwo Jima stand out as the most celebrated Marine Corps amphibious operations.

ARMY AMPHIBIOUS TRAINING

Hachinohe, Japan was the site of an Army regimental-sized amphibious training base with the usual wooden mockup towers firmly anchored on terra firma—no water, no waves. The soldiers learned to climb down the rope nets burdened with pack and rifle. It demanded good physical conditioning and coordination. Lacking was the experience of climbing down a rope net from an APA troopship rolling in the swells and entering a platoon-sized Infantry Landing boat that was rising and falling with each wave—a far different experience.

Later, the troops were ferried out to a quiet bay where training troop ships were anchored. There the practice was closer to the real thing, but still not up to a par with landing from open oceans.

When higher command decided that the troops were sufficiently trained, the entire Infantry regiment boarded APA Navy troop ships in the harbor. A giant flotilla of Navy transports, destroyers, destroyer escorts, and airplanes set in motion an actual amphibious operation on a larger and more realistic scale. The landing would take place at or near Chigasaki on Central Honshu.

As the flotilla/convoy sailed south, soldiers were given instructions for their mission upon landing. Mockups of the targeted beach were displayed with specific assignments for each squad, platoon, and company upon hitting the beach.

ARRIVAL OFFSHORE AT CHIGASAKI BEACH

The Naval Convoy arrived in the gray dawn as the Infantrymen assembled on deck in preparation to climb down the rope nets to the platoon-sized landing craft. The shoreline was visible in the far distance. The APA troop ships were rolling in the swells while the smaller landing craft waiting below the nets were bouncing with every wave and swell. The troops were carrying packs, rifles, canteens, cartridge belts, bayonets, entrenching tools, and other assorted gear, everything necessary for survival on the beaches. The weight was considerable for soldiers on dry land, let alone climbing down rope nets from the troopships into flat-bottom Infantry landing craft bouncing up and down in turbulent water. In a move that would have been humorous were it not so ironic, each soldier was required to have a life jacket. It was purely psychological because the life jackets were incapable of providing floatation to a soldier with all the gear he was carrying.

As the troops climbed over the side of the mother ship on to the rope nets, they experienced the difficulty of transferring into the assault landing boats. Three soldiers in the landing craft tried to hold the nets tightly in place, so soldiers would not be shaken off into the sea. That could cause death by drowning, as the life jackets were totally useless under the circumstances. But another problem confronted the men as they climbed down the rope nets. The small landing craft rose and fell with each wave as much as two or three feet. As the soldiers would near the bottom of the net, a wave would hit before they stepped into the boat, rising the boat high enough to knock them off their feet and into the bottom of the boat. The reverse could happen where the boat suddenly dropped several feet just as the soldier was releasing his hold on the nets, thus dropping him several feet into the bottom of the landing craft with pack, rifle and all. Needless to say, there were some wrenched backs and twisted ankles and knees. Thank God this was a training experience and not a combat landing. In past combat amphibious operations, many small Infantry landing craft were destroyed before reaching the shore. Further irony of amphibious landings: highly trained infantrymen died before they touched land or fired a shot. One of the sad fortunes of war.

LENGTHY RENDEZVOUS AND SEASICKNESS

Once all the troops were offloaded, the groups of landing craft made dozens of circles at sea, waiting for the signal to form lines, after which wave after wave of assault boats would head for shore at full speed. In the meantime, the constant wave action and swells caused the small boats to heave up and down. Within a few minutes, the 40 or so men in each boat were seasick and heaving their guts into their steel helmets before throwing the vomit overboard. Yet, in that condition of seasickness, they were supposed to be ready to charge off the boats once they hit the beaches. The hapless troops were cold and seasick but at least, they were not going to face live fire in this training mission

As the little boats rose on the waves, only sky would be visible to the troops. The next moment the boats could be in the trough of the wave action, and the crouched troops could see only a ring of water. Hopefully, they would be able to charge ashore even though debilitated by their sea legs and retching stomachs.

THE SERGEANT DISAPPEARED!

Signal flares were fired, the circling Infantry landing boats formed assault waves, and they raced to the shores of the targeted beaches.

Hitting the beach in one of the landing craft was another issue to be experienced. The craft hit the beach with a jolt and the ramp was let down. In the movies, we see troops charging directly off the front of the ramp. In a real landing, troops are instructed to jump off the sides of the ramp, as the swirling wave action can cause the ramp to ride up and over the exiting troops, hitting them in the back of the legs and pinning them under water, where they can be drowned or crushed to death.

The platoon sergeant gave the good old Infantry cry, "Follow me." He jumped into the waves and promptly disappeared underwater. The next soldier, seeing him disappear, refused to leave the landing craft. The other troops, not realizing that something was dreadfully wrong, were shouting, "Move out."

As the wave subsided, the first trooper observed the top of the sergeant's helmet under water. The sergeant's arms flailed desperately. His rifle in one hand, he tried to paddle toward shore with the other. So the second man backed into the shouting soldiers as far as possible and took a running leap into the ocean. He landed up to his chin in the water and felt the outgoing swell pulling him in deeper. Fortunately, he was able to struggle to the beach, where wet and exhausted soldiers were supposedly ready to do combat. The half-drowned sergeant finally made it to the beach and collapsed, alive but gasping.

As the platoon regrouped and looked around, nothing matched the mockups of their training on the mother ships. They had landed on the wrong beach due to the currents that had carried the whole flotilla down the coast while waiting for the landing signal. They landed on a bathing beach, with comely Japanese girls and their boyfriends enjoying the summer. In Japan, nudity wasn't frowned upon as in the States. Co-educational baths and other situations indicated that nudity was not an issue in Japan in certain circumstances. Imagine how thousands of G.I.s reacted, going from the trauma of a practice amphibious beach landing to finding themselves in the presence of pretty nude girls. The experience diverted their attention from the mission at hand, which was already confused because they'd landed on an unfamiliar beach.

Officers immediately tried to restore a sense of direction and mission by having the regiment move inland. But that created another problem. Troops carry two canteens of water when hitting the beach, because beaches are hot and dry, almost desert-like in the brilliant sun. Thirst is an issue until potable water is found or supplied. As the troops moved inland, some found themselves in a field with ripe and juicy watermelons. Needless to say, bayonets came out and before long the troops were feasting on the red ripe melons, satisfying their sweet tooth and thirst.

Once again, high authority was mightily concerned because the American army does not live off the land, seizing the produce and property of hapless farmers and civilians, as the "Red" armies did. We heard long after that event that the U.S. Army of Occupation had to make restitution to the Japanese farmer for his lost crop.

Eventually the regiment marched to a dirt road, where convoys of 2 ½ ton trucks were waiting to transport the soldiers high into the mountains through Gotemba and Yamanaka to Camp Mc Nair on the slopes the Japan's most famous mountain, Mt. Fujiyama.

After intense training, the regiment was ordered to Yokohama where another U. S. Navy convoy would transport the regiment to Inchon, Korea, where the beaches had been secured in previous actions, thus ending their preparation for future amphibious action.

THE FLOATING DRYDOCKS OF WORLD WAR II. BOATSWAIN 2ND CLASS ORVILLE EDWARDS

*I*n compiling these stories and events in the lives of the American servicemen and women, stories are sometimes interrupted by the death of the veteran, and the story has to be completed from historical data and the service person's DD 214, known as the separation document, which is given to all service people upon their discharge. The DD 214 is a miniaturized history of each individual and is an important document verifying military service. Such was the case of Boatswain 2nd Class Orville Edwards, who served in the South Pacific during World War II. The ribbons listed on his DD 214 certify his contributions to the war effort. The DD 214 shows that he served in the

Asiatic-Pacific Campaign, the American Campaign, and the American Defense. He was awarded the Good Conduct Medal and the World War II Victory Medal.

    Boatswain Edwards' experience was unique in the history of World War II in the Pacific. To a large extent, the war in the Pacific was much more of a Navy war than the war in Europe. Some of the greatest battles were naval engagements such as the Battle of Midway and the Battle of the Coral Sea. Guadalcanal was as much a Navy fight as a Marine and Army fight. Our battleship fleet was damaged or destroyed at Pearl Harbor initially. This left American sea power at a great disadvantage compared to the Imperial Japanese fleet. Fortunately, the nation's aircraft carriers and submarine forces were spared. Contrary to some of the great naval doctrines of that era, the battleship was not to be the principal fighting force in the great battles in the Pacific. Submarines and Naval airpower became the primary means for taking the fight to the enemy. The great Battle of Midway tilted the sea power back to the American forces with the destruction of Japan's major aircraft carriers.

    Outgunned and out maneuvered, the Japanese turned to one of the most bizarre approaches in modern warfare, the suicidal manned aircraft that were flown into American warships with devastating effect—the kamikaze.

    It was imperative that damaged ships be repaired and sent back into action as soon as possible. To tow them back to the continental United States or Hawaii was a long and tedious solution. The answer was the creation of "floating dry docks" that could repair damaged battleships and destroyers in the very combat zones where they could be returned to action quickly. Boatswain Edwards was one of those highly trained Navy technicians who could rebuild the external structure of ships and deck equipment. Part of this personal history was given in interviews and some was reconstructed from his DD 214 discharge.

BOATSWAIN ORVILLE "ED" EDWARDS' LIFE AND EXPERIENCES

Ed was born in Tucumcari, New Mexico. The family later moved to Maywood, California where as a young man he worked in a grocery warehouse. He relocated to Ventura, California when he was inducted into the Navy. Ed was 18 at the time of induction, typical of the age of enlisted servicemen. He listed his primary occupation as a student. Records show that Ed served for almost three years in the Navy and most of that time, he was deployed in the South Pacific. Typical of servicemen during World War II, once deployed overseas, service people remained either until the war was over or their service ended because of wounds or death. Sailors killed in action during those perilous times were buried at sea. World War II tours of duty necessitated long absences from family and friends and the salubrious benefits of normal civilian life and living. Every aspect of normal life was put on hold including education, careers, and family. Everything was for the convenience of the government. Ed's individuality was completely submerged: the war mission took precedence over Ed's very life.

ASSIGNMENT TO THE FLOATING DRY DOCK

The military often enlists callow and untrained youth with little or no experience other than high school. But, the exigencies of war require immediate skill development depending on the branch of service and the mission of any particular unit. Ed was thrust into such an experience, where he learned many skills. He was assigned to Tiburon, California where he worked on a floating dry dock being assembled for service in the South Pacific. It was a huge structure and was designed to repair ships that were damaged by Japanese air action and submarines. It was a novel concept and required much advanced engineering and training of naval personnel.

Youngsters in the military are given much more responsibility more quickly than in civilian education and industry—that is, if they accept the challenges of learning and leadership. Ed rose to the rank of Boatswain 2nd Class, which is equivalent to being a sergeant in the army. His particular assignment was to supervise any activity related to deck and boat seamanship. His rating required him to be knowledgeable and responsible for maintenance of a ship's external structure and deck equipment. Since the floating dry dock was designed to repair ships in the combat zones of the South Pacific, Ed had a secondary assignment to take charge of deck guns and crews when under attack. After an attack, the sailors were trained in damage-control parties to replace and repair the ship and equipment as necessary to continue with their primary mission. Boatswain 2nd Class Edwards held five specialized skill ratings.

TO THE SOUTH PACIFIC

Once *USS ABSD-2* was constructed in Tiburon, California, in the sheltered waters in the San Francisco Bay area, it was towed in sections to the island of Espiritu de Santos in the New Hebrides and later moved to the Island of Manus in the Admiralty Islands just north of New Guinea. The floating dry dock was so huge that it had to be floated there in sections before being reassembled. *USS ABSD-2* was so gargantuan that it could accommodate and repair one battleship and two destroyers simultaneously.

Once the dry dock was in place and reassembled, the work of repair could begin. But it was not ordinary work, for the damaged naval vessels came in various states of impairment. Some of the work included removing the bodies and body parts of sailors who had been killed and were still entombed in the bowels of the warships. That was a gruesome task, especially in dealing with the decay and corruption of bodies that had been exposed to the heat and humidity of that tropical climate. Sailors who performed the body removal details often commented that the stench of death never left those ships. Next, the repair crews had to remove the damaged innards of the ships, remanufacturing and replacing decks, bulkheads, gun turrets, and any other parts of the ship that needed replacement.

While *USS ABSD -2* was in the South Pacific, three destroyers were repaired: *USS Claxton, USS Killen,* and *USS Canberra.* In addition the battleship *USS Iowa* was repaired. These ships had been damaged in naval battles from Japanese bombs and Kamikaze suicide planes. Unexploded ammunition and fuel had to be removed before the dry dock crews could begin reconstruction.

For the modern civilian who reads Boatswain 2nd Class Edwards' experiences as a young man, it's difficult to fathom the depth and intensity of such a wartime experience. Like most of the World War II veterans, Ed returned home to begin a normal life. Despite the traumatic experiences and intense responsibility thrust upon Boatswain 2nd Class Edwards, there was no PTSD recognition. He went home to begin a life that had been put on hold for three long and lonely years. He married his sweetheart, Lillian, and, as was typical of men and women at that time, they had a long and lasting marriage of 61 years. That says something about the sense of commitment of these forgotten heroes from the "Greatest Generation." The Edwards family had three children, seven grandchildren, and four great grandchildren.

As a civilian, Mr. Edwards never forgot the debt this country owes to veterans who served so that we might have freedom from foreign tyranny. He supported the veteran's groups and was a member of the Veterans of Foreign Wars and the American Legion.

Mr. Edwards went to work at the military air station at Point Mugu in civil service, where he continued to contribute to the national defense efforts during the Cold War. Upon retiring, he traveled throughout the desert areas of the West where he was given the nickname of "Desert Rat." By the grace of God, his early wartime experience had prepared him for life making a contribution to the "Common Good" of this nation and providing for a large and growing family. Later, his one son, Randy Edwards, served with the American Infantry Division during the Vietnam War against the communist encroachment in Southeast Asia.

A CONSCIENTIOUS OBJECTOR GOES TO WAR

The Story of Paul Tenbrink

Conscientious objectors are people who eschew war and killing. They will not participate in military activities that require them to kill another human being. The most notable group of conscientious objectors are Quakers. Most other objectors come from various Christian religions such as the Jehovah's Witnesses, which are well known for their opposition to war and human government.

Various objectors approach the issue of war, killing, and their involvement with the military in different ways. Some objectors belong to no organized religion and express their opposition to war in political activism. Some objectors simply flee to another country such as Canada, as was a common occurrence during the Vietnam Era when the draft was in place. At that time, most young men were subject to induction into military service, with high numbers of draftees assigned to combat Army Infantry units.

If an objector was from an established religious denomination, he could be deferred if his activities and participation with his church were evident and substantiated. There were cases where some men tried to avoid the draft by becoming a member of one of these religious groups when they received their draft notice. It was hard to substantiate that they were truly bona fide objectors on religious and moral grounds and not simply draft dodgers who were trying to avoid military service. During earlier war periods, such men were often sent to prison.

Some conscientious objectors changed their views and became active participants in warfare. President Nixon was from a Quaker background and served in the active military during World War II as a Navy Pilot and commissioned officer. Sgt. York, a noted combat hero of World War I, was another conscientious objector who struggled with his basic religious beliefs about war and killing. He ended his World War I service with the Congressional Medal of Honor.

This story is about a conscientious objector who came from a bona fide religious background that did not object to military service in the cause of the nation, but for reasons of conscience, members could not participate in the killing of another human being. Many of these objectors were trained in humanitarian military activities such as the medical corps or as medics in combat units. This is a brief account of Paul Tenbrink and his military service in the South Pacific in World War II.

PAUL TENBRINK AND THE ARMY AMPHIBIOUS COMMAND

Our family was not particularly religious even though we had attended churches; however, at the age of 14, I began attending the

A CONSCIENTIOUS OBJECTOR GOES TO WAR | 113

Seventh-day Adventist Church in Washington State, along with my father. From that time forward, I have been a lifelong Adventist. I attended junior high school at a Seventh-day Adventist boarding school. The school offered a class called "Medical Cadet Training" to prepare us for military life, hopefully as medics.

Later I was called to active duty and passed the physical examination for Basic Training. I was classed as a "Conscientious Objector" and trained to be a frontline combat medic. Our church called us "conscientious cooperators."

One encounter with military authority occurred when I was in formation standing at Parade Rest and joking with the other soldiers. An officer approached me and accused me of acting like a high school kid. I snapped back, "I am a high school kid." The officer said nothing more but from that point, I was the "kid" in the regiment during training at Fort Lewis, Washington.

MEDICAL TRAINING AND MY CONFRONTATION

While in medical training we studied the Army's book of rules and regulations. One rule was that no army personnel can be refused when requesting to see the company commander. At that time, my First Sergeant with his six stripes had great authority and he used it. Because I had a question for our company commander, I went to the Sergeant's office and said, "May I." He immediately cut me off and gave me a severe tongue lashing as he shouted at me, "No!" He didn't even let me speak. It was part of his First Sergeant's mannerism and controlling behavior. It was consistent with his personality. I turned around and went to my next class.

Later I met the company commander. I saluted him and said, "Sir, I have just been refused to see you." He asked me a couple of questions and then I went on to class. One half hour later, I was called to the First Sergeant's office, where I received the worst bawling out of my life. The next morning we had a new First Sergeant. I don't know what happened, but the old First Sergeant was seen getting on a train with no stripes.

A PRACTICE PATIENT, A REAL MEDICAL ISSUE

We were on bivouac out in the fields. Part of the training was to use healthy soldiers as practice casualties for training purposes. I was designated to have an injury to my jaw. The X rays showed that I actually had two wisdom teeth that were coming through horizontally. There was only one dentist available. He proceeded to chisel out the wisdom teeth without any pain prevention.

SLOW BOAT TO THE PHILIPPINES

As was true with all troop transportation by ship, the open waters of the Pacific Ocean were subject to submarine warfare, and slow-moving transports filled with replacement soldiers made good targets for enemy forces. Troop transports are lightly armed and relatively defenseless against submarines. But we arrived in the Philippines without incident, except for my chiseled jaw, which was still swollen from the dental procedure.

THE SECOND AMPHIBIOUS ENGINEER BRIGADE

The Second Amphibious Engineering Brigade was an Army unit scattered throughout the Pacific and equipped with speed boats mounted with machine guns to hit small Japanese-held harbors. This activity helped divert enemy forces from Army and Marine amphibious landings.

The Second Amphibious Engineering Brigade had participated in the assault on Leyte in the Philippines in 1944. When I arrived, the Brigade was preparing for the attack and the landing in Japan, which was anticipated to be the final strike against the main Japanese Islands. Estimates of potential casualties, both Japanese and American, were projected in the millions. Fortunately, the atomic bombs at Hiroshima and Nagasaki ended the will of the fanatical Japanese military to resist any further. The war ended, to the relief of all soldiers and civilians who were facing an encounter of unimaginable death and destruction.

JUNGLE ROT

The Philippines has a hot and humid climate, and I was afflicted with "jungle rot," a condition caused by microorganisms and microbacteria found in tropical climates. Jungle rot can start with a simple scratch that creates a sore which, if untreated, can affect muscles, tendons, and bones. In severe cases, it can lead to amputation.

Much of the fighting in the South Pacific subjected troops to hot, humid climates where troops were exposed to this disease. I contracted jungle rot, and had to be hospitalized. After the war ended, I was put on one of the first hospital ships returning to the United States and a cooler climate. While awaiting discharge in the United States, I was married. We eventually had three children. Six months after returning home, I was discharged.

LIFE AFTER SERVICE

I attended Walla Walla Seventh-day Adventist College and pursued majors in religion and education. After college I worked at Boeing

Aircraft for a number of years, expediting lost orders and preparing purchase orders. I visited Ventura, California while on vacation and loved the area. I met a man who had a landscape and gardening business but was having difficulty keeping up with the work because of ill health. He offered me a partnership in the business. I accepted and moved to Ventura.

I enjoyed the work and took classes at Ventura Community College in subjects such as horticulture. Later I was offered a job at Loma Linda Medical Facility, where I took care of the grounds with 20 employees. I was there for ten years before I was hired by the University of Michigan in the same capacity. The Michigan assignment had more to do with snow removal than what I was used to in California. Eventually I returned to California, and lived in Glendale. There I continued my work at the Adventist Medical Center where I was in charge of the grounds and landscaping. Upon retirement, I returned to the Ventura area and Ojai Valley.

I am an active member of the Ojai Valley Veterans of Foreign Wars and active in mission projects and work in Indonesia, working with Pastor Paul Emerson of the Seventh-day Adventist Church.

THE PREOCCUPIED ARMIES OF OCCUPATION

Health and Moral Issues of the Army of Occupation

There is a side to an occupation army that many service men and women wish to forget or suppress, especially for many of those who occupied foreign countries after World War II. The closest that any author has ever come to the subject is Michener in his book, *Sayonara*. But Michener makes a point of emphasizing the romance and glamour with a forward-looking political correctness in interracial relationships. A few salient passages describe the real encounters of G.I.s with the occupied population.

America of the 1930s and 1940s was still a very puritan country in terms of morals and mores, but it was the puritanism inherent in the immigrant people from the Christian nations of Europe. The churches were strong and well attended. In fact, it was a rare family that did not have strong connections with the church. Sex was a taboo subject and rarely mentioned in polite society. News media and radio broadcasts were sanitized. The French attitude toward sex was viewed as a symptom of a degenerate society. Abortion was a nonissue and if it did occur, it was underground. Birth control was a hit or miss situation. Pictures of women in bathing suits, called pin-up girls, were about as risqué as anyone dared display. Usually, these pictures were displayed in some off-the-track section of a garage or warehouse where women would not see them. It was a man's thing. The whole topic of sex was hidden, and seldom left the gym locker room or the little groups of worldly-wise older boys. Gossip and scandal did occur, often blown out of proportion to any real event.

In general, society was straight, morally clean, and had a strict standard of conduct. Young men and women married, had families, took them to church, and stayed married to one person for their lifetimes. That was the norm. The churches all had youth programs that centered on the teachings of the church and the Holy Bible. Generally, youth activities such as movies, dances, picnics, or special youth oriented activities followed the religious instruction.

The word "gay" meant happy and light-hearted. The word queer just meant that the person was odd. Homosexuality was a crime, a felony. Such people, when caught, were imprisoned, and dark rumors occasionally circulated about someone suspected of being too feminine in mannerism. The subject was never understood. Lesbians were unheard of for the most part. It was a culture of innocence mixed with great curiosity and an underground discussion of human sexuality from a bawdy and often illiterate sharing of misconceptions.

THE MILITARY EDUCATION

The military is a great educator and leveler of humanity. The soldier encompassed the best of humanity and the most curious, but it was hard to tell who was righteous and who was truly profane as long as the army remained in the United States where the contact and influence of family, church, and peers exerted a strong influence on behavior. Of course there were those more experienced soldiers who were continually trying to shock their buddies with exaggerated tales of masculine accomplishment, introducing ideas that the sheltered G.I.s never dreamed of.

A few fellow G.I.s, when returning from liberty, told many wild tales of their escapades. These immature fellows were particularly proud of "knocking" a girl up, which meant getting her pregnant. The married men were above such talk and looked upon these braggarts as immature punks with no grace and no couth. The majority of men reflected the more disciplined and puritan ideal of the existing society. They attended chapel, read their Bibles, and respected women in general. They missed their families, since this was a time when communication was often not instantaneous.

Shipment to Japan was an uneventful experience. Thousands of soldiers were crammed into the cargo holds of the Liberty or Victory ships that plied the North Pacific ferrying men and war materials to the Far East.

Our first encounter with the Japanese people was upon landing in Yokohama and boarding trains for Hachinohe in Northern Honshu. I was loaded down with my field pack and other gear when I saw an old Japanese lady overburdened with bundles trying to open the door to the train station. As awkward as it was, I rushed forward to open the door for her, as any American gentleman would do. The poor woman was surprised, flustered and embarrassed that a man would do such a thing. She ran and hid some distance from me. I was perplexed until one of the old timers explained that in the Orient, there was no such thing as chivalry toward women. I soon learned that the role and plight of Oriental women was extremely low. They were like slaves to the men, and would tolerate great abuse unflinchingly.

Within a week, we were billeted in a tent city outside the town called Hachinohe. On one of my first weekends in the camp, I heard a ruckus in the tent across the company street. I heard a woman cackling and laughing and a lot of whooping and hollering from the men. How the woman got into the camp was a mystery but she wasn't there for any good. From that point on, I heard unbelievable tales of the Japanese sex trade in the local town. It seems that fishing, rice farming, and pimping were the main industries of the local population. As subservient women, dutiful and obedient, they were enlisted in the sex business. For Americans, it was shocking to have husbands rent their wives and daughters for sex, or sons pimping for their mothers, or brothers selling their sisters. If you have ever been in a foreign country in Europe, Asia or Africa, you may have experienced aggressive sex marketing.

Soldiers on leave were barraged with men and women plying their trades or selling their wares. To the soldiers, new to these Asiatic shores, and with strong memories of American women in their mind, and for those who retained a sense of chivalry, the whole sex scene was repulsive.

It was not uncommon for two girls to start a tug of war over each G.I. One of the most famous popular songs of the day was, *Come On-a My House, My House, Come On.* That was the exact sales pitch of these street girls, called Geishas, but who weren't Geishas at all. The patter of their broken English sales pitch was, "You G.I. San – Come on to my house. You likee vely much. Want you want?

Short time? Long time? Joto ichiban josan." Translated: short time was a quickie of a few minutes with a price of three dollars. Long time was overnight and was more expensive. Joto ichiban josan meant that she was an okay number one girl. Needless to say, the morals and mores of Japan were a shock to the puritan sensibilities of both Catholic and Protestant soldiers. But as time went on, the foreign appearance diminished, or was no longer noticed, and the G.I.s began to see beauty in some of them and would comment that they looked almost like the girls back home.

Needless to say, the soldiers became preoccupied with attracting the young Japanese females' attention. A second lesson about this part of the Orient was that nudity was no big deal. Their public bathing houses were often filled with naked people, young and old. Nobody so much as flinched except the uninitiated soldiers. Ultimately, G.I.s stripped and joined the party like natives.

Mt. Fuji and upslope from Yamanaka Lake and the town of Yamanaka became the next training location. This town was small, with houses on either side of the main dirt road leading to another town of Yoshida. It was on the shores of a beautiful lake surrounded by rice paddies. The main industry by day was farming. It was a common sight to see the Japanese farmer hitch his wife and daughter to the plough. While he pushed, they pulled the plough through the rice paddy. Early in the morning, these same women carried two buckets on the end of a long pole. Stopping at each house, they collected the "night soil," or feces and urine. Quickly, they trotted out to the field where they sloshed it into the rice paddies prior to plowing. Needless to say, all fruits and vegetables grown in or on the ground were contaminated with the diseases from the human waste. Americans cringe at the thought of walking in such mulch and human filth, but to the Japanese farmers and farm girls, it was a vital component of life and survival in a land short of resources.

The main industry in Yamanaka by day was farming such as described in the previous paragraph. At night, the industry was pimping and whoring by the same farmers and their women. This town had been exposed to this lucrative trade for some time, since Fujiyama was one of the main military training bases.

The regiments soon moved into the five hundred tents on the slopes of Fujiyama, and that night they were able to relax.

The third lesson of the uninitiated, hormone-driven soldiers was that bad behavior had an additional price. Suddenly, syphilis, gonorrhea, crabs, and other more exotic venereal diseases began to take their toll. The old timers who were well acquainted with the dangers of uncontrolled sexual activity were least affected. They knew how to protect themselves. The good little Sunday school boys, who hadn't a clue what was going on, soon found that they were in deep medical trouble. One poor lad hid his diseased condition until his private parts were so swollen that he was unable to make revelry one morning. When the sergeant went into the tent to investigate, the lad was writhing in agony. Medics came with stretchers and carried him away. We never saw or heard from him again. It was then we heard about Oriental strains of VD that were so bad that the afflicted G.I.s were sent to special facilities in the Orient to be warehoused. That was rumor and hearsay, and never confirmed. But we did wonder about that poor fellow.

The biggest scandal of the base was that the head Japanese barber for the camp was also the head communist for the Prefecture of Yoshida, where we were encamped and in training for combat in Korea. Military intelligence discovered that the communists were bringing diseased girls into the local towns to deliberately infect the soldiers. Whether our information was totally correct or not was not important, many soldiers of the regimental combat team had contracted VD. That was a serious threat to the fighting capabilities of the unit.

The fight against VD began in earnest. The Chaplain Corps was enlisted to give the troops morality talks. We were to remember our wives, our sweethearts, and our Christian upbringing. The religious appeal was calculated to bring the fallen soldiers back to the puritan standard of sexual abstinence. But it didn't work. The next line of defense was to issue condoms and prophylactic kits for each soldier before he was given his liberty pass. Soldiers with higher standards would refuse them and so would be denied the liberty pass. Many condoms ended up being used for rubber bands to blouse the boots since rubber bands were unavailable.

The religious approach through chaplain talks and appeals to decency and self-control failed miserably for several reasons. Thousands of young men with their raging hormones are difficult to restrain, especially when in the "candy store" of uninhibited sexual opportunity. The influences of home, church, peers and the social and moral restraints of a puritanical society were nonexistent. There was a pervasive attitude that everyone was doing "it." Sexual expression was so free and easy that it didn't take long for conventional morality to take a vacation. The moral leadership and examples were not in place.

Even more alarming was a Christmas party where the chaplains were invited along with the other regimental officers. Only two chaplains did not attend because they were aware that drunkenness would be standard fare for this party. The two Chaplains announced that the birth of Jesus Christ should not be celebrated in such a manner. Of course, they were considered party poopers, but they were accorded a high respect for taking a stand consistent with their position as chaplains. The other chaplains were good fellows and got as drunk as the rest of the soldiers. Soldiers witnessed a chaplain in a drunken stupor trying to get his cup refilled with gin. As the bartender—also drunk—tried to fill it, the chaplain's status and moral authority was diminished forever. The chaplain was one of the boys, but he was not a real chaplain in the eyes of the troops.

Many of the family men remained faithful and loyal to their wives. It was a struggle, but they diverted their sexual drives into other activities, which kept them occupied. Some climbed Mt. Fuji, some went on weekend tours to Kofu, or Atami, or the pachinko parlors. English films with Japanese subtitles were available in the larger towns like Nagoya or Numazu.

These G.I.s remained true to their wedding vows, but it was not infrequent, after their men's long absences from home, that some of the wives and girlfriends found other interests. The "Dear John" letter was the most painful and demoralizing experience of many soldiers. While in Japan, it was not uncommon for husbands jilted in this manner to go to town and abandon all restraint in one horrible debacle of drunkenness. Later, in combat, soldiers who received such letters were inclined to become indifferent to the dangers of combat operations. They felt betrayed, confused, and deeply depressed. While still in the states on a Saturday afternoon relaxing in the barracks, we heard a shot. A shot in the barracks area was very out of place. We rushed outside to see soldiers running into the Command Post. One of our lieutenants had received one of those notorious "Dear John" letters. He walked into the Charge of Quarters and asked him if he had any 45-caliber pistol ammunition. The CQ, trying to be helpful, said, "No Sir, but I have carbine ammo." The lieutenant took the carbine ammo, went into the office and closed the door. He lay down on the bunk, placed the business end of the carbine into his mouth, and blew his brains out.

There were many G.I.s who tried to follow their religious and moral standards, but there were an equal number of soldiers with less lofty ideals who were committed to make all their comrades into their own sordid image and likeness. If you didn't drink, they had to get you drunk. If you didn't smoke, they had to get you into the habit. These G.I.s were most pleased if they could get a good Christian soldier to "get laid." That was their expression for sex. They had a particular name for these soldiers. When the "cherry boys" succumbed to temptation, the worldly soldiers were in high glee, like they had scored a winning touchdown.

Some of these setups were during liberty, when the G.I.s were least suspecting. After a long ride to a town on the trucks, hot and dusty and tired from the long weeks of maneuvers, it was not uncommon to just go to the quaint hotel room and sleep and sleep. The rustling of a kimono and the sliding of the wood frame doors might awaken the G.I. to see a beautiful young girl kneeling in front of him trying to make conversation. The girl could be underage and innocent. It could be mere curiosity and a desired friendship.

Ultimately, soldiers whose recreation centered on partying in the local towns surrounding the huge military bases in the Orient began to view all Japanese women as corrupt. Like towns adjacent to military camps in the USA, soldier towns in other parts of the world are a magnet for corrupt people who are looking for what appears to them as easy money, no matter the personal human degradation. The lives of such people can be short-lived in those circumstances, which can often lead to disease and violence.

A lovely Japanese woman was not immune to the wrongful attentions of misbehaving soldiers.

One day, a train was traveling from Tokyo to Yoshida when a beautiful American Japanese woman boarded. A lustful soldier immediately began pestering her for her services and attempted to negotiate a price. She politely and firmly rebuffed him. That only angered him into worse insults and crudity. The Japanese passengers ignored what was happening, pretending not to see. Other soldiers on the train assumed that it was as it appeared. They assumed the woman was a higher-class practitioner of the oldest profession who was holding out for greater gain or bigger fish. But the beautiful young woman was becoming angry. Her perfect English should have given a clue to the soldier that there was something different about her. The soldier was too brutish to understand until she laid it out plain and simple. She stated, " I am an American. I work for GHQ in Tokyo and my boss is General so and so. If you do not leave me alone, I want your name, rank, serial number, and the name of your commanding officer." The red-faced soldier beat a hasty retreat. The assumption that because she was a Japanese woman she was not a proper woman was totally mistaken.

Once the soldiers were far removed from the soldier towns, it was soon learned that the Japanese were a different people than commonly assumed by the soldier denizens of the so-called geisha houses. In fact, most soldiers did not realize that a geisha was a highly trained entertainer, not a synonym for a street person. Away from the camps, the Japanese people were friendly, courteous, and decent people, generous to a fault and very hospitable.

One tragic event that affected many servicemen was that lonely soldiers, usually teenagers who never had girlfriends in the States, confused love with raging hormones. Unprepared for the Japanese sex trade, many became so involved with a particular woman that they proposed marriage. An engagement ring was often a class ring. For the Japanese girl, this was a golden opportunity to latch on to paradise, for they had their misconceptions about America and Americans. They thought that all Americans were rich and rolling in money. Michener got it right when he described this scenario in his book on the subject. Commanding officers and chaplains along with the bureaucratic apparatus of the Far East Command sought to discourage these liaisons.

Returning home on troop ships, soldiers were still billeted down in the cargo holds of the merchant ships one level above the bilge water. Up on deck there were cabins for the newly married soldiers and their "Japanese war brides." It was an embarrassment to watch them. The Japanese girls who had been plow girls in the day and involved in illicit activities at night were now dressed in fashionable American garb with short skirts instead of kimonos. They sat on the deck in groups on their hunches like they did in the fields after emptying their honey buckets. It was most entertaining for the cruder soldiers to stroll the decks leering at these confused women out of their cultural element displaying themselves unwittingly. The kimono does a better job protecting decency than a short skirt. When a G.I.-San saw his wife being the butt of sordid jokes about her anatomy, he rushed to his spouse, telling her to stand up. He spoke English. She only understood Japanese. The cultural clash was evident big time. Some of the marriages continued as life-long commitments, but some were disasters. It depended on the intelligence of the woman and the tolerance and compassion of the man.

Through the fifty plus years since these events, the men and boys returned home, married, had families and established careers. Most left with a sense of decency, fell from the standards of their upbringing, but reverted to the values and standards in which they were raised. As I communicate with many of my old comrades, I never cease to be impressed with the quality of their lives and their commitment to the values of their early youth before wartime service. The wheel had gone full circle.

THE THUNDERBIRD FROM OKLAHOMA

The Story of Thomas Flowers, Army Artillery NCO/ Naval Officer

Oklahoma, in the center of the country, is the home of many Indian tribes that were shunted there by the United States government as a policy of relocating Native American people. Once Oklahoma was opened for settlement, it wasn't long *before there was an amalgamation of European and Indian blood. Indian culture and symbols abound throughout the state, including the Oklahoma National Guard with the 45th Infantry Division. Their first symbol was a swastika. Most people don't know that the swastika was an American Indian symbol. After the rise of the Nazis in Germany, when they appropriated the swastika as their symbol from the ancient Aryan people, the American Infantry Division had to change to another Indian symbol. They chose the Thunderbird. This is the unique story of a veteran of that division who later changed branches of service and became a Naval Officer.*

Lt. Cmdr. Flowers' medals and decorations.

MY BEGINNINGS

My early family life was simple and hardscrabble. As a young child, I was sent to live with my grandmother who survived on the economic edge. After all, Oklahoma was in the center of the Dust Bowl of the 1930s, which created disastrous living conditions. It was the era of the Great Depression, which affected the lives of all who lived through those tough times. Assisting my grandmother was a Black lady who I looked upon with great affection, for she was a kind and gentle woman with a heart for a young child who was bereft of the normal family relationships.

I grew up quickly, and it was not surprising that at the age of 15, I joined the Oklahoma National Guard. It was not legal and my mother did not sign for me. She asked how that was possible and I told her that I lied about my age and the recruiting NCO asked no questions. She asked me what birthday I gave. When I told her, she mentioned that she was not married at the time I gave. She asked, "And do you know what makes you?" I thought about it and realized that I was born on the other side of the "blanket" at the age I had given. My father split when I was three years old. I did not get along with my mother's husband at that time and when he mistreated her when I was older, I was ready to "clean his clock."

For me, the Thunderbirds became a home away from home. I was assigned to a gun crew for the 105 mm towed howitzers. As time progressed in National Guard training I was advanced as an NCO and became quite proficient in laying a battery, fire direction, communication, and transportation. As artillerymen know, the slogan of the Artillery is "Shoot, Move, and Communicate." We trained at the national home of the United States artillery at Fort Sill, Oklahoma.

CALL TO ACTIVE DUTY

The Korean War erupted early in June of 1950 and the soft easy life of the U. S. Army on occupation duty in Japan and Germany had not prepared the post-World-War-II soldiers for the rigors of combat. Our politicians had assumed the United States did not need a large, well-trained conventional military because conventional wars were a thing of the past. In contrast, the Communist nations had ramped up their military forces with modern equipment and well-trained men. Needless to say, when President Truman committed our poorly trained and equipped occupation army to Korea, the nation was shocked as reports from the battlefield described our poorly trained soldiers being overrun and units so decimated that they almost ceased to exist. Weapons and ammunition were in short supply and the armories of the National Guard throughout the nation were scoured for crew-served weapons. Soldiers in basic training, some who knew very little about weapons and combat, were hurled into battle almost sacrificially as the nation suddenly ramped up the draft and intensified training.

September 1, 1950, a desperate nation called up two National Guard Divisions, the 45th Oklahoma National Guard Thunderbirds and the California 40th Infantry Division known as the Sunburst. On paper, it appeared that these divisions were more ready for service than other National Guard Divisions.

Like the regular army, the Guard units had their own equipment, training and supply problems. Yet, my division had a fine cadre of experienced World War II officers and Noncoms. They knew the ropes and how to train troops. Initially, these Guard divisions had two missions, training and defense of the continental United States that was left exposed as every available soldier was being shipped as replacements to our battered forces in Korea.

Sgt. 1st Class Thomas Flowers in Korea with the 45th Thunderbird Division

Sending the Guard overseas was considered a political hot potato, but the troops in Korea had left Japan relatively undefended against a Soviet threat. It was then that my division was shipped to Hokkaido, the northernmost Japanese Island at that time. Hokkaido was far north and had a cold climate and bitter winters, but that is where we trained, a training environment not unlike what we would experience in Korea. We trained as individuals, as gun sections, as gun batteries, and eventually as regimental combat teams, a very high level of training for any military organization. Once again, the political thought was that Guard divisions would never enter combat. But the battle-weary and worn down regular army divisions were in desperate need of retraining and reorganization to restore fighting effectiveness.

THUNDERBIRDS REPLACE THE IST CAVALRY

Sgt. 1st Class Thomas Flowers with 105 mm Howitzer Gun Crew in Korea

The 1st Cavalry Division had been in the war almost from the beginning. They had suffered horrific casualties in constant combat, first against the North Koreans and later against the Chinese. They had fought from the Pusan Perimeter and the Naktong Bulge, where our army almost faced annihilation. They were in the breakout coordinated with MacArthur's Inchon landing. They raced toward the Yalu River on the Manchurian border thinking the war was over. They took a heavy hit when the Red Chinese Army enveloped them and almost destroyed their fighting ability. Thrown back to the 38th Parallel, they regrouped and settled into a seesaw battle with static lines, trench warfare, and combat patrols. It was at that point that the 45th Infantry Division was ordered to Korea ,sailing from Hokkaido in Northern Japan to Inchon, Korea. From there we were trucked to the front lines to replace the 1st Cavalry Division.

THUNDERBIRDS THUNDER DOWN ON THE HAPLESS COMMUNISTS

Regular army units had a jaundiced view of National Guard Divisions. I was with C Battery of the 171st Field Artillery. I was a nineteen-year-old Sergeant First Class. My gun crews moved into position to take over the howitzers and fire missions. A grizzled middle-aged Sergeant First Class of 1st Cavalry battery said, "I'm sure glad to see you guys." He then asked me how long I'd been at this, meaning as an artilleryman. I replied, "Since I was fifteen." I said, "We'll take over your next fire mission." He offered his advice or assistance and was ready to offer any help or pointers on manning the howitzers. I responded that we knew the drill and were ready as we took over the gun position. We took over with a military professionalism that astounded the veteran soldiers. The 1st Cavalry Sergeant First Class watched us go into action and after observing the proficiency of the Thunderbird gun crew commented, " I have never seen a better trained group in my life!" He had just assumed that we had been weekend warriors, not aggressive, highly trained, and motivated soldiers. We commenced firing. With the guns registered and targets located, my battery settled in to the routine of fire missions as called by the forward observers on the front line and in the infantry outposts.

In time and with replacements, the unit was changed from the original soldiers who had trained together for so long before entering combat. One day, having been with the forward observer team with the infantry, I'd just come back from the front. I was relaxing that night having a cigarette on top of a bunker. Across the dirt road, a searchlight unit had moved in. (Little known to people unfamiliar with the Korean War, searchlight units were used on cloudy nights to light up the battlefield for the defending infantrymen. The light was very effective in helping the men in the trenches observe the enemy's movements in front of them. But, the searchlights drew artillery and mortar fire.) Soon rounds began to fall in C Battery's firing position. I ducked into the bunker when the phone lines began calling for a fire mission to silence the enemy artillery. Rounds were falling on C Battery even as the fire mission was ordered. The section chief refused to leave the safety of the bunker. I picked up the phone and went out the bunker ramp. Three howitzer crewmembers followed me out. We received range and deflection settings. We commenced firing and silenced the enemy artillery, thus protecting the "moonbeam" searchlight unit as well as C Battery.

THE SORRY STATE OF EQUIPMENT

One wrinkle in this switching of position was that we had to take over the 1st Cav. Equipment, which was in very sorry condition and had to be repaired and reworked. That included howitzers, communication equipment, and vehicles. However, with almost one and a half years of superior training, the Oklahoma Thunderbirds demonstrated that we were one of the finest fighting units ever to serve in the Korean War.

A LASTING IMPRESSION

Every soldier in combat faces death and eternity. We destroy the enemy, and we see the faces of those our forces have destroyed. But when we come face to face with seeing the destruction of our own comrades in arms, we never forget. We know that except for the grace of God, we could lose life or limb. For those who suffer, they can lose all hope for a future. One of my most disturbing experiences that haunts my memory even today was seeing trucks piled with slain Oklahoma Thunderbirds with arms and legs hanging over the sides. As a soldier I knew that those who serve can be ordered to fulfill missions from which we will never return alive. That is war! For those who face peril and death, they are never more alive than in those circumstances.

Though an artilleryman, I was ordered to serve with the Forward Observer Team directing fire into communist positions. Though an officer usually heads the Forward Observer Team, the enlisted men often conduct the fire missions. One troublesome position was an artillery piece hidden in a cave that was periodically moved out into a firing position, fired, and then withdrawn into the cave. One of my most exciting calls as a forward observer was directing fire at this position. I called in the coordinates on the cave. I made the right call and one of our rounds entered the cave, probably ricocheting into the interior where it set off secondary explosions. I knew I had a hit that sent many communist to the "happy hunting ground," an old Indian expression. After all, I am part Cherokee!

I CAME HOME

I survived the carnage of battle with the mixed emotions of pride in service and the knowledge that I did my duty as a combat, frontline soldier. Yet, forever etched in my memory is the understanding of what it means to be a soldier, and the price free men pay for the good of the nation.

As with most returning Korean War veterans, my thoughts turned to my academic future and with the help of the limited G.I. Bill I entered college while working in various jobs including in the oil fields as a driller—a hard, dirty and dangerous job. I worked evening shifts and went to school until I earned my business B.S. from East Central State College. I entered the Air Force Reserve with the thought that I would make the Air Force a career, but Air Force Officer training did not allow married trainees. By that time I was married and had one child on the way, so I enlisted in the Navy where they allowed married personnel. I went to preflight school in Pensacola, Florida. There I learned to spell "airplane" and trained in T-34 propeller-driven planes. But military downsizing after Korea terminated my flight training. The Navy didn't want to release me and assigned me to Aircraft Maintenance Officers School in Memphis, Tennessee. Later, I was assigned to the Naval Intelligence School in Norfolk, Virginia. I went there with a young family in tow. After that, I was assigned to Lemoore Naval Air Station in California, followed by a tour in Hawaii.

ASSIGNMENT TO THE PHILIPPINES

From aircraft maintenance, I was assigned as a young Ensign to the Naval Intelligence Staff, investigating criminal activity and doing background checks. My family came with me and my wife Carol had a good time socializing and mixing with the Filipino wives and base personnel and their families. Some wives complained about the putative hardships, but my wife was a real trooper and turned the deployment into an opportunity to get to know the Filipino people person to person.

I was authorized to carry a pistol because of hazards of my assignment, but I decided that if I accidentally had an encounter and shot a Filipino national, I could be prosecuted under Filipino Law. My partner Hector was a Filipino attorney and he did carry a weapon. Generally, we did our investigations in plain clothes. One day, there was a report that a sailor had been stabbed by a woman prostitute, and we had to drive to the location to investigate. While driving there, we were pulled over by a Filipino policeman who reported that an American sailor was causing a problem among the Filipino people in Pasay City, a suburb of Manila.

We located the house where the troublesome sailor was holed up. It was a house of prostitution. I approached the front door and my Filipino partner went to the back door. He was armed, I wasn't! I started to knock on the front door and I had a thought that I should stand to the side as I knocked. I knocked and announced that we were from the U. S. Navy Intelligence. Bang, bang! Two shots came through the door where I would have been standing. I didn't know the sailor was armed and dangerous. The suspect ran out the back door. Hector hollered, "Tom, are you okay?" I yelled in the affirmative as I ran around to the back where Hector had a pistol pointed to the head of the troublesome sailor. Hector asked if I was really okay. I replied, "I wasn't so much afraid as I was very angry at the sailor for shooting at me."

We took the bad guy and turned him over to the Marines at the American Embassy. We knew that he wouldn't get any slack from the Marines, who had little tolerance for misbehavior, especially in a foreign country abusing the citizens. With that little incident resolved, we went on with our investigation of the stabbing, our primary investigative focus. It was an altercation between a "business woman, a Filipina national" and the sailor. Young sailors don't always understand the psyche and temperament of foreign nationals.

A FLOATING BOMB

Naval Lt. Cmdr. Flowers on carrier duty

Warships are floating warehouses of explosives and volatile fuel storage. Fire and explosions are a constant danger and attention to safety is never-ending. A following assignment was to an aircraft carrier, probably one of the most hazardous warships afloat, loaded with munitions, rockets, bombs, aviation fuel, and other inflammables. As a safety officer, I was responsible for safety issues.

Naval Officer Tom Flowers on Carrier Deck off Vietnam

One day I was in the aircraft hydraulic workshop below the flight deck when I noticed a hydraulic actuator was connected unsafely to a test stand and outside the parameters of safety requirements. A burst hydraulic line could spray atomized and explosive fluid throughout the workshop. An electrical spark could cause an explosion. I pointed this out to the officer in charge. He exploded verbally and told me that it was his shop and that he was in charge. Of course, Navy language can be colorful and not too polite. I was told to bug out in no uncertain terms. I turned and walked out.

When I was about 30 feet away there was a terrific explosion with a ball of fire coming out of the hydraulics workshop. Five sailors were inside and had been burned almost beyond recognition. The concussion pushed me through an unlatched door and on to the deck. The Wing Commander came out of his office and directed me to go into the hydraulic shop to assist with the rescue. Our destroyer screening vessels had seen the explosion and subsequent fire and were pumping water into the hanger deck. The burned sailors were pulled out unconscious or semiconscious. The officer I had warned was lying on the floor in sickbay. I knelt down beside him. He asked, "Is that you Tom?" He couldn't see. I answered in the affirmative. He asked, "How bad off am I?" I didn't want to say, as half of his face was badly injured. I quietly said, "You won't win a beauty contest." Hearsay later reported that he was never normal after that.

Los Angeles Reserve Police Officer Thomas Flowers in Uniform

I had been injured in the blast and had medical issues after that, but at least I was a whole man.

I served two tours with the Navy in Vietnam before retiring and returning home. With civilian life, I started a new career as an Airport Patrol Officer in Ventura County.

Retiring from county service, I became a Reserve Police in the Los Angeles Police Department where I was assigned to a detective and sex offender section working on cold-case investigations. I worked with a detective and served as a plain-clothes officer on the streets of West Los Angeles, staking out neighborhoods with high crime activity.

Several years before retiring from the Los Angeles Police Department, I was named Reserve Officer of the year.

Tom is typical of his generation, coming from struggle and hardship; yet, he served his family, nation, and communities over his lifetime, contributing to the "common good." But one impelling factor in Tom's personality was an abiding sense of a God-given faith throughout his life; he practiced the virtues of a Christian soldier. For many years, he was the principal lay leader of St. George's Anglican Church, where he served on the Vestry as Senior Warden. His roots were most humble. He was steeped in the teachings of his Christian Scripture, and so Tom led his life. Tom was blessed with a loving wife, children, and grandchildren. In the absence of the dire poverty in which he was raise, he achieved remarkable success in life.

144 | VETERANS' STORIES BOOK III

*Flowers' Family while Tom was with the U. S. Navy—
with Two Navy Comrades*

ATOMIC GUINEA PIGS, CONVENIENCE OF THE GOVERNMENT

■ ■ ■ ■ ■ ■ ■ ■ ■ ■

The experience of Corporal John Pressey RA 28107103 during the Atomic Bomb Tests – November, 1951- Yucca Flats, Nevada

The following material from the Department of Defense certifies Corporal John Pressey, Jr. as being a member of the United States Army who was involved in the atomic bomb tests in Nevada in 1951: "DEFENSE THREAT REDUCTION

AGENCY-DOD – 1/6/2010. Participation in Atmospheric Nuclear Tests conducted – VA criteria military records excerpts enclosed confirm your participation in Operation Buster/Jangle conducted at Nevada Test Site in 1951. October 10, 1951- November 6, 1951 – Camp Desert Rock at Indian Springs, Nevada in route to Fort Lewis, Washington. Document DTRA: Defense Threat Reduction Agency – Fact Sheet- Description of Operation; Buster/Jangle: at Nevada Test Proving Ground- Seven nuclear detonations: four detonations air dropped, three shots including one tower, one surface, and one underground detonation. Involved 11,000 Defense Department Personnel in observer programs. Intended to test nuclear devices for inclusion in the weapons arsenal."

BACKGROUND AS A COMBAT MEDIC IN KOREA

I enlisted in the United States Army in 1949 and was sent to Fort Ord, where I received Basic Training with the 4th Infantry Division before being sent to Osaka, Japan as part of the Army of Occupation. My assignment was with the 25th Infantry Division (Tropic Lightning) and the 27th Infantry Regiment (Wolfhounds). There I received training as a medic attached to an Infantry battalion. In June of 1950, the unit was deployed to Korea with the understanding that there was some kind of disturbance in Korea and we may face armored North Korea jeeps. Our initial contact in a blocking position between Taegu and Daejeon exposed us to a North Korean attack from T-34 Soviet tanks and North Korean soldiers. That was the beginning of twelve months of heavy combat, first with the North Korean Communist Army, when we were pushed back to the Naktong Bulge. With the Inchon landing, we broke out of the Naktong Perimeter, moving north beyond the North Korean capital of Pyongyang as we raced toward the Yalu River. Unknown to most of the troops in the field in the far north of North Korea, Chinese armies had entered the war and enveloped much of the Eighth Army, creating a new war and a new crisis for the American combat soldiers.

MADIGAN ARMY HOSPITAL—STATESIDE DUTY

I served throughout those desperate times for twelve months, witnessing horrific casualties, being wounded myself, returning to my unit as a combat medic, being decorated for heroism, and eventually rotating back to the United States in August of 1951 and assigned to the 374th Convalescent Center at Fort Lewis, Washington. While assigned to the 374th Convalescent Center, an Army Reserve Unit called to active duty, we went through classes on first aid and medical familiarization training, a quirk and silliness of the army system. We were being trained after the fact. I'd served months in the field in Japan and twelve months as a combat medic with a rifle company in Korea.

WE GET TO SEE LAS VEGAS?

In November, the 374th boarded trains at Fort Lewis for transport to Las Vegas, Nevada. Upon arriving in Las Vegas, we were transferred to trucks and driven about sixty miles north to a place called Yucca Flats, an area designated as a test site for Atomic weapons and atomic experiments. Initially, our unit served setting up tents and digging latrines for the VIPs. It was bitterly cold at night in our squad tents, so much so that we broke out shelter halves and ponchos to cover ourselves in addition to three wool army blankets. Yet the days were so hot that the heat was almost insufferable, especially in contrast to the nights. While we set up the VIP tents, the winds were ferocious. Tent stakes had to be driven in quite deep to keep the tents from being blown away. The commander of this operation was General Keene who had commanded my 25th Infantry Division in Korea.

EVERYTHING IS SECRET

Everything that happened or was about to happen was classified. We were briefed about not discussing anything pertaining to the activities and events at Yucca Flats. Our unit was called together for an "orientation." We were told that we would get to see the dropping of an atomic... We were trucked approximately twenty miles outside of camp where we were told that we would be able to witness a B-52 drop an atomic bomb. Wow! A free fireworks show! A military commentator stood on a raised platform. He had special goggles; however, the troops had no special equipment. We were told to sit on the ground and face away from the blast. We were told that it was perfectly safe and we were beyond any danger of radiation, but if we thought there might be a problem, we should take a shower afterward and that would take care of any radiation problem. Supposedly, we were quite a distance from the pending blast or "ground zero." The commentator reminded us to face away from the blast site, and told us to close our eyes and sit in a crouched position with our head between our legs.

THE FRIGHTENING BLAST SO CLOSE

When the bomb was dropped, I felt a powerful hot wind hitting my backside from the blast. After the hot wind struck us, we were told to turn around so we could see the mushrooming cloud. Another wind came from the opposite direction as the towering red mushroom cloud sucked back the air blast into the rising cloud. The blast seemed awfully close, despite the statements of the military commentator. I could feel the ground shaking from the concussion of the detonation. The whole event seemed close because of the huge size of the fireball, the heated wind, and the gigantic mushroom cloud. The army inspectors declared that the event was safe and harmless. Yet, we were forbidden to talk about what we saw and experienced. We were briefed about the classified nature of the event. From the "observation site" we were trucked approximately twenty miles back to camp. Security was very tight. Later while in the camp, I witnessed another atomic bomb detonation, but no precautions were taken because it was assumed that the camp and personnel were too far away to cause any untoward effects.

All told, we spent about 30 days out at Yucca Flats. We spent the time watching movies and drinking green beer, which made some of us sick. We were allowed to go to Las Vegas for one Sunday, where the civilians had more information about what was happening at Yucca Flats than we did. Some of the soldiers just got drunk before returning to camp Sunday evening.

BACK TO THE 371th EVACUATION HOSPITAL

Upon leaving Yucca Flats, we were trucked back to Las Vegas where we boarded trains for the trip back to Fort Lewis, Washington. There I was assigned to the 371 Evacuation Hospital Orthopedic Wards for the rest of my enlistment.

Looking back on the experience, I have a sense of irony after serving in twelve hard and dangerous months in the worst part of the Korean War, being wounded and decorated for heroism only to come home sans parades or public displays of appreciation. Instead, I was used as a guinea pig for the good of the service. Many people died prematurely from the effects of the radiation including civilians in St. George, Utah who were downwind from the explosions.

15

A LIFE INTERRUPTED — A DRAFTEE'S DILEMMA

■ ■ ■ ■ ■ ■ ■ ■ ■ ■

The Story of Edwin Marks, Artilleryman

The Korean War was not anticipated by President Truman and the Pentagon Planners. It was considered obvious by them that the possession of atomic weapons was a sufficient deterrent to conventional warfare, in which masses of troops engaged in conventional battles. The Truman administration allowed the dissipation of our monumental military presence of World War II until our Armed Forces were but a shadow of the once mighty air armadas, vast naval presence on all oceans, and the millions of men in our fighting units. Smugly, the military planners and politicians assumed that "the Bomb" was the answer to military preparedness. In contrast, the Soviet Union, a communist dictatorship, began an unprecedented buildup of conventional armaments. They increased espionage in order to obtain the atomic bomb secrets of the United States, and began an unparalleled subversion of the free nations of the world, including "home grown" communist sympathizers and agents within our own country.

Our military forces had devolved into third-world-level "paper tigers" undertrained and poorly equipped, with no new weapons, equipment, or modernization programs. Our fighter planes and bombers were sent to the "bone yards' in the Arizona deserts to be scrapped. Many of our proud naval vessels were used for atomic bomb tests in the South Pacific and sunk. Artillery pieces, tanks, trucks, guns and all the accoutrements of modern warfare were jettisoned into the depths of the oceans—long before environmental concerns were commonplace.

While the United States demobilized, the Soviet Union extended its hegemony over all of Eastern Europe. Communist insurrections broke out in Latin American, the Middle East, and Southeast Asia. China succumbed to the "Red Menace' in 1949. Gradually, the Western World was being overwhelmed by communist subversion, conquest, intimidation, and ruthless rule.

Korea was one area where the Soviet Union was given a foothold in the Far East. Most Americans did not have a clue about Korea. It was small and obscure, certainly not a player among the major nations of the earth. Yet, the Soviet Union, which did not fight the Japanese in World War II until the atomic bomb was dropped, savagely attacked Manchuria, and took over Sakhalin Island and the Kuriles. The Soviets were allowed to occupy the northern half of Korea above the 38th Parallel. They installed a puppet regime under a Soviet Citizen named Kim Il-sung, who became a Communist dictator. Korea below the 38th Parallel was under the influence of the United States and the South

Korean government. South Korea was woefully neglected by the Truman administration, and he declared in so many words that South Korea was not in the strategic interest of the United States. That was the trigger for the invasion of South Korea by a massive Soviet-trained and equipped army infused with the ruthlessness of Asian armies, armies that lacked humane consideration for the vanquished and defeated. The Geneva Convention did not apply. Compassion for a suffering humanity was a weakness in their eyes.

An unprepared America committed our untrained skeleton Japanese Occupation army piecemeal into a fight no one fully understood. Equipment was lacking and back in the United States National Guard Armories were being stripped of mortars, machine guns, recoilless rifles, and any other crew-served weaponry. The plight of our hapless soldiers in Korea was brutal beyond comprehension. Fortunately, American air power had not been fully degraded in the Post World War II stand down.

Suddenly, the industrial might of the United States and a military buildup became paramount in the Truman administration. The manpower that had just missed the draft in World War II was tapped and called to service. Many of these draftees were older men in their mid-twenties who were just beginning careers, starting families, and accepting the responsibilities of mature men. With the draft, their lives, their marriages, their careers, and family interactions were put on hold. For over 54,000, their life was not just on hold, their life was over. More than 103,000 would return home blind, crippled, deaf, or with incapacitating wounds so that they would never again be able to enjoy life to its fullest. Many would come home with the guilt, fear, and terror common in the battlefield. Why guilt? Because comrades perished while they survived. But how does one explain that to civilians?

Ed Marks was one of those young men preparing for a career, marriage, a family and the good life. He had not planned to enter the service. In fact, when a number of his buddies joined the local artillery battery of the California National Guard, Ed wanted nothing to do with such a commitment, not because he objected to the military, but because he had so many other plans for his future. When the Korean War broke out and that National Guard unit was mobilized, Ed was thankful that he was not part of that activation. He was a civilian with a future. But that was not the end of Ed's story!

ED'S STORY IN HIS OWN WORDS

I was born in 1929 in Colorado. My father was a deliveryman for a bakery and had several bakery routes. However, Colorado is cold country in the winter, with ferocious storms. Colorado experienced a particularly terrible winter in 1936 with extreme cold and continual snow. Estes Park, snow covered, was closed to all traffic. It was then and there that my father decided to sell his bakery routes and move to California, a state with a mild climate and more salubrious weather. We had relatives in Ventura County in the city of Ventura. It was there we settled and I attended Sheridan Way Elementary School, Cabrillo Junior High School, and Ventura High School and Junior College. I was interested in boats and cars. I planned to become a forest ranger. I had met my future wife, Jackie, who was attending Redlands College studying to become a public school teacher.

As a kid, I started to go to midget car races in Denver, Colorado back in 1942. In Ventura, I had a buddy who had a roadster. I decided I would like to have one also so I went to work to earn money. I found an old 1932 pickup truck frame in a drainage ditch in Somis. The farmer said to take it since he didn't want it. He also had the body and bed of the truck in his barn that he wanted to get rid of. I worked in the Ford garage after school and the parts manager sold me parts at cost. We built the pickup truck and finished it in about June of 1947. We worked in a garage off the alley in back of my parent's house. My Dad helped and my Mom fed a bunch of us guys a couple of times a week.

In the meantime, I bought a 1936 Ford Coupe and fixed it up to go to the drag races at Saugus.

One of the special things we did was go to the dry lakes at El Mirage east of Palmdale. We went to see how things were run and to race some cars. Next, we took a ten-man tent to use as a garage and as sleeping quarters. We were able to take our tools and welding equipment if we needed to do repairs.

I had many different work assignments and experiences while attending high school and junior college. I worked as a parts manager for a Plymouth/DeSoto agency and in management at a gas station. By this time I had done a lot of things to the pickup truck engine so I could race it up in Lompoc and Santa Maria, California. I started dating my future wife, Jackie, who was attending Ventura Junior College, and took her to one of the races in Lompoc.

Later, we became interested in boats and water skiing.

THE BIG PLANS AND BIGGER INTERRUPTION

Jackie transferred to Redlands University about the time I received my draft notice. I was told to report to Fort Irwin out in the desert, not too far from Redlands University. By then, Jackie and I were planning to be married. We would still have time to be together on weekends, but one of the ironies of fate was that I was assigned to an artillery battery in the same National Guard division that my buddies had joined before the Korean War. Another irony was that shortly after I was sent to Fort Irwin, most convenient for Jackie and me, the artillery battery was transferred to Camp Cooke near Lompoc where I had been racing cars. Not so convenient for Jackie and me.

The 140th Field Artillery antiaircraft unit was from San Diego originally and had been activated with the 40th Infantry Division for Federal service. Again, this was an ironic situation for me because back in Ventura, I thought I had the last laugh on my buddies who had joined the National Guard only to be activated a short time later. I thought, "Wow, I'm sure glad I hadn't joined the National Guard." But fate was against me and I was drafted. And where I was assigned was the same division recently activated! Yet, there was a silver lining to that assignment related to my many youthful experiences and employment.

While in the 9th Grade, I was employed as a driver for a doughnut bakery, driving donuts and rolls out to the oilfield workers in Ventura. I made the grand sum of $12 per week for a job I obtained by falsifying my age so I could drive. I was only fifteen. My Dad worked for Prosser's Bakery while in Ventura and with my teenage work experience in the donut shop and bakery, I had some familiarity with food handling.

My experience with motor vehicles would give me a leg up on the less experienced draftees. That was probably the reason I was assigned to the 140th Field Artillery Anti-Aircraft Unit as a driver of a Quad-fifty caliber Half-track vehicle. It was a highly mobile unit, and drivers had to pay special attention to maps, compass readings, and directions for driving as well as firing. The Half-track vehicle could be used against enemy aircraft, or direct or indirect fire support of the Infantry.

In addition to driving, I was assigned to instruct recruits on the use of the lensatic compass, and on map reading with grids and coordinates, azimuths, and location of various field operations and maneuvers using military maps.

As a highly mobile unit, we moved often to different firing ranges to give us more experience for actual combat operations.

Quad 50 Half-track used for anti-aircraft fire and direct infantry support

GUARDSMEN'S ENLISTMENT PROBLEMS

Many of the Guardsmen had enlistment problems. Some were too old. Some enlisted illegally underage. Some were not physically fit. Some had dependent problems that created hardships. As a draftee, that gave me an opportunity for advancement. The mess sergeant had six dependents and couldn't continue to serve because of this hardship. Knowing my civilian work experience and background, the Captain asked me to be his mess sergeant. That was a real break because I could advance in rank to Sergeant First Class and better pay.

POLITICS AND THE NATIONAL GUARD UNITS

Guard units were much more sensitive to state politics, and the popular rumor was that Guard units were merely to bolster homeland defense and that they would never go overseas. The draftees began to take comfort in that rumor, until we were ordered to Camp Stoneman in the San Francisco Bay area for deployment to Japan to replace regular army units now fighting in Korea. Japan was relatively defenseless in the event of a Soviet attack. After all, we were in the Cold War with Russia, and anything could happen. So we boarded the aging troop ships from World War II at Fort Mason for a two-week cruise to Japan.

Troop ships are not luxury liners. We were crowded by the thousands in the cargo holds of the ship. Conditions were primitive and confining. Though boring, it was a matter of enduring the conditions of shipboard life until we reached Yokohama, Japan. Once again, the political rumors maintained that the 40th Infantry Division would never fight in Korea because of stateside politics.

The Guardsmen that stayed with the division were of two classes. The first group was made up of officers and top ranked NCOs who had served in World War II, many in combat including our division generals, Eaton and Hudelson. They were experienced men who provided excellent leadership. Opportunities opened for new recruits to advance and move up the ranks quite quickly. The second group of Guardsmen were youngsters in the 17-19-year-old bracket who wanted to remain with the activated Guard. The more capable and motivated soon adapted, and with training and commitment, they too had opportunities for leadership.

For the time being, the two National Guard Units, the California 40th Division and the Oklahoma 45th Divisions, were protecting Japan from Soviet attack, and were thrust into the most rigorous and comprehensive combat training that any army division had experienced since World War II. Training included amphibious landings, all live-fire regimental maneuvers, coordinated tactical air–ground assaults, tank /infantry attacks, live overhead artillery fire, flame throwers, demolitions, ski trooper training and endless field maneuver in the worst of weather and climatic conditions.

YOU WILL SEE COMBAT IN KOREA

Back at Camp Cooke on February 28, 1951, our commanding general had massed 20,000 men of the Division on a parade field. He had addressed us stating, "No matter what you have heard, this Division, the 40th Infantry Division, will fight on the field of battle in Korea." He warned us to take our training seriously and not ignore the training and lessons that may save our lives. At that point, we knew deep down that the politicians were wrong. We would be fighting in Korea.

The word came sometime in late December of 1951 that the 40th would deploy to Korea. Shortly, we were loading our half-tracks and trucks on LST's in Yokohama as a huge Naval flotilla was assembled to sail to Inchon, Korea. A soft snow was falling, the first snow of the winter season in Japan.

FIRST IMPRESSIONS LANDING AT INCHON

I had experienced cold in Colorado as a child, but Inchon seemed much colder. The temperature was minus 15 degrees. There was also a big difference from the cold in Colorado: in Korea, we were living outdoors, in tents, and in the fields.

We didn't have the luxury of a warm house or permanent shelter. Frostbite was a clear and present danger for all of us. The greatest comfort was body heat conserved by three layers of socks and shoepacks. We were layered with shorts, thermal long johns, wool pants, snow pants, T-shirt, wool shirt, faux fur lined jackets, M-65 field jackets and trench coats. We had both gloves and one-finger mittens for firing a rifle.

The civilian population was hungry and cold. They were pitiful. Inchon and Seoul had been bombed out and burned out. It was a shock to see the conditions of the poor Korean civilians. It saddened us.

Upon landing, we moved inland and north about 50 miles to our first positions only to be greeted by incoming artillery and mortar fire. The Reds saw us coming. We had to move back out of range on the dirt roads. From the time we took our positions, we had very limited contact with civilians.

A winter in Korea, 1951-52.

But delivering hot meals to combat units on or near the front lines was our job. All too often, mess trucks and our track vehicles were spotted by Chinese forward observers who zeroed in on our mess trucks and our Quad-fifty Half-track vehicles. We lost vehicles to enemy fire in these operations.

A LIFE INTERRUPTED — A DRAFTEE'S DILEMMA | 161

Korean chogee boy delivers food at the entrance to a bunker

As mess sergeant, the first thing I did upon moving into a secure position was to set up the mess tent. As they say, "An army moves on its stomach."

My mess had a reputation for good food, and many ranking officers would come to our mess for pies and other mouthwatering goodies. When my mess needed supplies, they were told to give me whatever I needed and not to say anything. We had Koreans helping with the mess and for their services; we gave them the bones and leftovers from the cooking. They were hungry people with no visible Korean government support. They were essentially on their own.

The weapons of the unit were exposed to extreme cold and the soldiers had to be careful to wipe off any congealed oil on the moving parts of their weapons less they jam and fail to fire. Though

I was not on the guns, we had to be ready for any contingency. That required our weapons to be ready at all times.

My cooks and I were thankful for the warmth of the heating and cooking.

RELIEVING THE 24TH INFANTRY DIVISION

Mess tent for the artillery unit

The 40th Division relieved the 24th Infantry Division that had had a year and a half of hard fighting. As the war wore on, that unit had to survive by "on-the-job" training as fillers replaced the dead, the wounded, and the rotated soldiers. There was a need to regroup and retrain. But there were problems.

There was a well-established contempt of the Regular Army brass for National Guard units. The irony, once again, was that most of the men of the 40th Division were not National Guardsmen but draftees and regular army soldiers. Of course most of the ranking officers and NCOs had been Guardsmen, but many of these leaders had served in active duty regular army units during World War II. The Army brass had it all wrong.

The equipment of the 40th Division was mostly used equipment from World War II but it was well maintained, and the men of the 40th kept all equipment in excellent condition. Upon our arrival in Korea, the Army brass did something so ruthless and so vicious that it jeopardized the lives of every 40th Division soldier—they made us trade our good serviceable equipment for the worn out, useless equipment of the 24th Infantry Division, so they could train with our equipment. That included crew-served weapons, trucks, artillery pieces, etc.

But the men of the 40th improvised. They sent home for spare parts for the trucks. They cannibalized until they had reequipped the division with serviceable weapons, trucks, and other equipment. That is a testimony of the quality of training and leadership of the men of the 40th. Whether draftee, regular army enlisted man, or National Guardsman, there was pride in service.

HOT CHOW FOR THE TROOPS

It was deemed good for the morale of troops to have hot chow, so positions were selected hidden from enemy view where the troops could be fed; however, it was not uncommon for Chinese artillery and mortar fire to zero in on these hidden spots, destroying mess trucks and even our Quad fifty Half-track vehicles that supported the infantry with indirect fire.

EMERGENCY LEAVE

Several months after landing in Korea, I received word that my father was desperately ill and I was given emergency leave. After my leave, I was sent to Camp Roberts where there were many soldiers returning from Korea and they were just waiting for discharge papers. Soon I would receive my papers and head for home.

Upon separation from service, I resumed my civilian life with one difference. I had that pride of service derived from my military training and experiences. I supported all the local veterans groups and events from the Massing of the Colors, Memorial Day in Libby Park, the Korean War Veterans Association, the Veterans of Foreign Wars, and the Christmas veteran's remembrance dinner.

I was married while still in the Army in 1951. After my service, I went to work as a Parts and Service manager for Weber and Cooper in Ventura, and moved to Oak View in 1958. There we raised our children. Like I had done with my father, my son and I, David, built up a 1957 Chevy and raced it at Irwindale. In 1975, we bought a 22-foot Schiada racing boat and we joined the Offshore Power Boat Racing Association. We raced in the ocean in Ventura and Marina del Rey where I blew an engine—and that ended my boat racing. My interests in cars, racing, and my work experiences all contributed to a successful tour of duty in Japan and Korea. Like most veterans, I will never forget those days.

I have a good wife, a good family, a successful career and I honor the service of all comrade veterans to this day, knowing that many comrade veterans paid the ultimate price for our liberties.

Ed Marks at the 40th Infantry Division Memorial at Vandenberg AFB/Camp Cook.

A Korean War vintage tank on display at the 40th Division Korean War Memorial at Camp Cooke, California. (Now Vandenberg Air Force Base.)

16

A SCREAMING EAGLE IN VIETNAM

The Story of "Doojie" Seliger

The Great depression was over. World War II had ended. The Korean War had erupted as the Cold War became a hot war. The Southeast corner of Los Angeles County was the industrial center for the California based factories surrounding several cities that housed the blue collar industrial workers of that time. South Gate had been mostly fields of mustard, wild oats, and gopher holes. It was an area undeveloped. During the 1930s, developers were selling residential lots for $50 each, with most potential buyers scoffing at such worthless land. WW II caused an influx of workers who rapidly bought homes and filled the burgeoning housing tracts of modest bungalows unique to Southern California.

Interspersed in this corner of Los Angeles County were farms, orchards, aircraft plants, automobile factories, steel and aluminum *mills, forge shops, foundries, the Firestone Tire Plant, Farmer John's meat packing plant, and the industries and scrap yards on Alameda Street which separated South Gate from Watts, a section of Los Angeles that was populated by colored people, as they were referred to then. The inhabitants of South Gate were mostly new transplants from the East and the Midwest. Many of the young workers fed the military needs for foot soldiers and trained manpower in the more technical branches of service, the Air Force and the Navy. It was a good community, with many opportunities for young people.*

The wider Los Angeles area offered numerous cultural, educational, and sports activities. The Los Angeles school system had excellent schools and colleges, museums, a planetarium, and beautiful parks such as Exposition Park and later the La Brea Tar Pits development. The coliseum and Wrigley Ball Field offered first class sporting events; and of course, Los Angeles was the center of the movie industry.

A GREAT BOYHOOD

Doojie Seliger was born in that environment and community. His Jewish father had recently returned from World War II after serving in the First Infantry Division known as the "Big Red One." A large "Red One" was the emblem on its particular shoulder patch. Doojie's mother was a Christian from Oklahoma and raised Doojie in that faith, a religion that Doojie still follows in active worship at the Church of the Living Christ in Ojai.

Doojie benefited from all the salubrious opportunities afford him during his boyhood. He was in the YMCA Indian Guides. He frequented South Gate Park, which was close to his home. He loved to be engaged in sports such as baseball and acquired skill at tennis, a sport where his father excelled. Doojie got to meet some of the tennis stars through his father. They included Bobby Riggs, Jack Kramer, and others. He participated in sports and visited the Los Angeles Tennis Club periodically.

Doojie's grandparents lived on the other side of Los Angeles near the La Brea Tar Pits, which he visited frequently while staying with them.

He marveled at the YMCA facilities where they had a track above the indoor basketball court. He thought that was unusual and creative.

When Doojie became a teenager, he was fascinated with autos and hotrods. That southeastern area of Los Angeles County became famous for these older cars redesigned with souped-up engines for drag racing and high performance racing contests. Doojie was part of that culture. Sometimes, the youth of South Gate High School engaged in illegal street racing. It was part of the bravado of the teenagers.

THE WATTS RIOTS—MY FIRST ENCOUNTER WITH ARMED CONFLICT

During the summer of 1965, an event occurred so alarming that it struck fear into the blue collar inhabitants of that section of Los Angeles County. In some ways, Los Angeles could be described as a quiet and undeveloped city prior to World War II. With the onset of World War II and the rise of defense industries throughout the Los Angeles Basin, there was a shortage of labor. Women were hired in droves, especially at the numerous aircraft plants manufacturing fighter planes and bombers. It was the era of "Rosie the Riveter." But the need for workers extended to the Southern states, and many Blacks migrated to Los Angeles for the blue collar jobs in defense industries. Many of these Blacks came from states that practiced "Jim Crow" discrimination against Blacks—separate drinking fountains, separate restaurants, separate travel facilities, and sometimes no access to basic human services.

Despite California being known as a progressive state, it was not for the Blacks! Housing was segregated, and restrictive housing codes kept Blacks to limited sections of town. The California, real estate Boards supported these so-called restrictive covenants. Fair housing laws did not exist for the most part. Then there was an issue of police misconduct. Blue collar workers were known to force Blacks out of their blue collar enclaves in incidents of racial conflict.

MINOR ARREST LEADS TO URBAN WARFARE

The trigger event in the riots was the arrest of a Black for drunk driving. What should have been a minor incident escalated into a riot as police were attacked and rumors increased the spread of the violence.

The riot that eventually turned into a civil war in the streets of Los Angeles became known as the "Watts Riots." It had been named after that section of town where there was a high concentration of Blacks, due to the restrictive housing rules in white areas. The riots involved many killings, and whole sections of South Central Los Angeles were destroyed by fire.

TEENAGER DOOJIE STANDS GUARD WITH A SHOTGUN

South Gate was a blue collar, working class, white enclave., Alameda Street and the railroad tracks separated South Gate from Watts.

I could see the huge clouds of black smoke as Watts burned and mobs of Blacks pillaged businesses and homes, fighting the police and even killing innocent victims. The Los Angeles Police Department could not contain the anarchy. The 40th Division of the California National Guard was mobilized with machine guns, bayonets, and live ammunition to quell the riots. My buddies and I joined 18 South Gate policemen at the border of South Gate and Watts. We entered the fray as a defense force, arriving in a Woody Station Wagon.

Someone had acquired weapons and for the first time in my life, I was standing with a shotgun in my hands ready to repel the rioters from Watts.

Military might coordinated with the police force eventually quelled the riots. Ironically, some of the National Guardsmen were Blacks from the Watts neighborhoods. Their concerns were for the safety of their families because the violence and destruction was confined mainly to the Colored section of Los Angeles.

SCHOOLS AND CLUBS

One impressive activity for me was membership in the DeMolay, a youth group sponsored by the Masons. There, I learned Robert's Rules of Order and parliamentary procedures. The De Molays emphasized patriotism, flag etiquette, and the seven precepts of good citizenship.

While in high school, I took print shop with the old style printing presses and moveable type which was often thrown at other students, much to the consternation of the teachers. It was a popular course for boys. Some of us also took home economics, not for the cooking, but for a chance to meet cute girls.

GREETINGS FROM UNCLE SAM

I graduated from high school in the winter of 1966. There were six to seven hundred students in that class. Most of the boys were ready for the draft as the war in Vietnam was heating up. I took a job at Douglas Aircraft where I worked as an assembler, a jig maker, and riveter. Because of my clearance, I became aware of some of the most advanced aircraft still on the drawing board, including the famous DC 10.

As the Vietnam War heated up, and because I was classified as 1A, the highest priority for induction into the Armed Forces, it wasn't long before I was on my way to Fort Ord, California for basic training. I completed Basic Training as an Infantryman, but several of my buddies wanted to join the "airborne." It's been said that you have to be crazy to jump out of a perfectly good airplane. Most paratroopers have heard this joke many times from their erstwhile acquaintances.

PARATROOPER IN THE VAUNTED 101ST AIRBORNE

Completion of training

Fort Benning, Georgia became my next abode, where after intense physical development and unlimited miles of running and jumping, I completed the five mandatory jumps to qualify me for my Parachutist Jump Badge.

Wartime parachutists are not recreational jumpers. Jumping into live-fire combat zones means that you have to have a parachute that is designed for a fast descent so as not to be caught helpless in the sky by enemy fire. The minimum jumping altitude is three hundred feet. Any miscalculation and if the chute does not open immediately, the trooper is facing certain death. Even when everything works properly, the parachutist hits the ground so hard that broken bones and other injuries are all too common. Military jumping can cause injuries that affect the joints many years after service. Normal jumping heights are about 450 feet which gives the parachutist a little higher margin of safety. Later, in Vietnam, parachuting was replaced by rappelling from helicopters, especially after the disastrous experience of the 173rd Airborne jump, which was very costly in lives lost.

WE ENTER VIETNAM

Soon after completion of Jump School, a brief three weeks furlough was enjoyed before we left by airplane for Vietnam, flying from Alaska to Japan, and finally landing in Cam Ranh Bay. It was early spring, 1968, but upon landing at Cam Ranh Bay at

Arrival in Vietnam

night, we were hit with the intense humidity of Vietnam—and the mosquitoes. We were then given a place to sleep in the sand next to plywood billets built by the Seabees. We stayed there until we were sent to Bien Hoa. We received more training that emphasized skills of observation and continual alertness, which would help us stay alive once we entered combat operations.

Shortly after we arrived, the encampment came under mortar fire. I hit the floor and pulled my mattress over me for protection. That was my first experience under enemy fire. In the morning I found shrapnel fragments on the ground.

ENLISTED MEN ARE ALWAYS WRONG

From Bien Hoa I was sent by plane to Hue and then by truck to the 101st Airborne. There it was decided whether I would do convoy duty or "hump"—go out in the field. Reporting to the Sergeant, I was told I was in the wrong area. I was sent by plane to another area where the Sergeant wanted to know why I was late. I tried to explain but he kept yelling at me that it was my fault, not the Army's. Of course, the Army is never wrong. Leadership is always right! And the hapless buck private is always guilty or wrong. It's his fault if mistakes are made. At that point, I was assigned as lightweight infantry.

The Sergeant told us that we were never to be caught walking. We had to run everywhere. We hated that sergeant at Camp Eagle. Soon we were sent out to the field in helicopters that landed in rice patties. We would encounter villages and engage in firefights as we attempted to neutralize enemy activity in the area. From there, we returned to Landing Zone Sally, which was a major base of operations.

NVA MOST DEADLY

In our field operations we encountered Viet Cong, who would open fire and disappear. When we encountered North Vietnamese soldiers, they would stand and fight. They were much more dangerous. While in the field we engaged in many close firefights with the North Vietnamese Army (NVA) and the guerilla Viet Cong.

I carried an M-16 rifle but was cross-trained to handle any assignment in the infantry platoon, including being a radio telephone operator and the use of an M60 heavy machine gun.

Vietcong on the march

Soldiers of other allied nations were involved in our operations. They included South Koreans, Australians, Turks and South Vietnamese soldiers known as ARVNs.

ISOLATED PATROLS IN JUNGLE WARFARE

For long periods of time, we were out in the field, the jungles and rice paddies, on ambush missions where we would set up ambushes at night. Other times we were on "tiger patrols." A tiger patrol would set up a temporary ambush before moving to another location, always seeking to engage enemy forces by stealth and cunning. We'd heard that the NVA had a Russian advisor with them and we were constantly trying to capture him, but we never did. While on reconnaissance patrols we were constantly looking for enemy caches of weapons, ammunition and food. We carried our rifle cleaning rods so we could probe any soft earth. When we hit something with a distinctive metal sound we would unearth the cache and remove or destroy its contents.

RTOS, THE LIFELINE FOR FIELD UNITS

I was an RTO (Radio/Telephone operator) for the platoon much of the time. We operated far from our bases. Communication and coordination for air drops were relayed by the RTO at the direction of the combat leaders. Of course RTO operators were a real target for the enemy, because loss of communication could result in the annihilation of the unit.

We relied on periodic helicopter drops for resupply of food, ammunition, and medical supplies, plus evacuation of the dead and injured. They were the workhorses of the unit.

SHEER PHYSICAL AND MENTAL STRAIN

We went weeks without bathing or a change of clothes. When we would return to our base of operation, we reeked of filth and sweat. Blood sucking leeches clung to our bodies. Our clothes could almost stand up by themselves they were so dirty. While out in the field, we would be alert and active most of the night. Only with dawn were we able to get a few hours of sleep. Living with constant danger that required alertness, we were often sleep deprived and exhausted. After five and a half months, I was so exhausted that I was almost unable to function physically, plus the psychological tension and fear that accompanied soldiers every minute of the day and night 24/7.

In the combat zone of Vietnam, we used a different technique for landing. Instead of parachute drops, we learned to rappel out of hovering helicopters. For the first five months, my specific unit was out in the field, not garrisoned in encampments. Along with food, medicine, and ammunition, helicopters also brought us replacements.

MY SCARIEST 4TH OF JULY

That night four experienced combat veterans took out 10 new replacements to teach them about setting up an ambush. As we left the concertina barbed wire we entered an area where a napalm charge had been set. Somehow, by accident, the napalm canister was ignited, spewing flaming gasoline over one of the patrol members. One soldier ran from the flaming gasoline tripping and busting a rib. Another man tripped knocking himself out. It was a scary fiasco but became even scarier when the Sally Landing Zone lit up along with the main base at Camp Eagle with massive explosions and gunfire. Even Hue was lit up with explosions and gunfire. We were receiving fifty-caliber fire from friendly forces. We thought that a major battle had erupted. It wasn't until later we learned that the troops at Sally LZ, Hue and Camp Eagle were just celebrating the 4th of July. With four injured men we eventually made it back to base, not knowing the whole experience was due to mistakes and friendly fire. I've heard that 20% of war casualties are from friendly fire.

Helicopters returning from assault

I have deliberately left out the more gruesome aspects of my experiences in Vietnam.

ASSIGNMENT TO THE S-4 SUPPLY

After five or six months of combat operations with all the attending dangers, fears, and brutality, I was assigned to the S-4 in charge of distributing rations. This assignment was far different from previous wars where lower class rations were available to the troops in the field. Class A rations were the highest quality, such as would be found in mess halls and permanent installations. Class B rations were largely powdered eggs, powdered milk, and other dried foods that were reconstituted by the cooks and mess stewards. For the troops in the field, they had mostly canned food and packets of salt, sugar, coffee etc. These rations were called C rations, and the soldiers were responsible for their preparation or eating under field conditions.

Supply truck returning to camp.

Vietnam saw food rations of a much higher quality and in more abundance than previous wars. I was in charge of distributing steaks, ice cream, milk, and all the other foodstuffs familiar to Americans at home. These items were unheard of in the battlefields of previous American wars.

RETURNING HOME—LIFE AFTER WAR

Upon arriving home, I had no civilian clothes for the first few days. I went up to Hollywood to take in the sights. I was in uniform and as I was crossing the boulevard when a man came by and spit on me. I was caught so off guard that I didn't know how to react. Later, I came under the verbal assault and castigation that so many veterans of Vietnam experienced when returning home. A segment of the population hated us for being soldiers. The irony of this is that many of the same people who blamed us for the war had voted for President Johnson, the supposed man of peace, who was largely responsible for promoting the war. Many combat veterans from that era are still very bitter about how they were treated.

I returned home to take advantage of the G.I. Bill of Education, attending East Los Angeles Community College where I earned an AA and an A.S. degree with a technical license in microbiology. At the same time I worked with my father in the roofing business, a business that was hard and hot, working with hot tar, tar mops, and heavy roofing. About that time my father, an entrepreneur, purchased a mobile home park in Lancaster, California, where I moved to become his manager and maintenance man for a number of years.

I had a very talented girlfriend who lived and worked in Ojai, California. I followed her there and together we worked developing a business screening logos on T-shirts and other clothing items.

After a divorce and the birth of my daughter, I became a handyman and community volunteer worker for many of the community events sponsored by civic and patriotic organizations, including the 4th of July, Memorial Day, Veterans of Foreign Wars activities, Wine Festival, Music Festival and offering assistance to handicapped and disabled persons. I am active in the Church of the Living Christ and assist with youth activities. I serve as the Commander of the Ojai Valley Veterans of Foreign Wars Post 11461.

THE BACKBONE OF NAVAL OPERATIONS—THE AIRCRAFT CARRIERS

The Story of Commander Les Allen

BENEFITS OF MILITARY SERVICE

Military service has its hazards, but overall the hazards to the vast majority of enlistees and officers are offset by the positive benefits of training, discipline, and order. How many youth have been drifting idly, engaged in meaningless activities? But, upon volunteering for service, for whatever reason or motivation, their lives take a new direction.

A concomitant benefit of the more technical branches of service, including the Air Force, the Navy, and Army Engineers, is that this technical knowledge helps young men and women obtain employment after service, utilizing their military skills in civilian jobs.

Military service often thrusts young people into positions of leadership and to the application of technical skills quite rapidly when compared to civilian life and civilian companies. Military training is focused and directed toward the needs of specific missions as compared to civilian institutions, such as colleges and public schools, where years may be spent taking courses that have no bearing on specific future needs. Generally, there is a purpose behind every aspect of military training. Training is direct, with minimal time wasted in peripheral studies that have no practical use.

Most honorably discharged veterans return to civilian life with a better sense of direction and focus. They are in a position to evaluate what is important and what is meaningless.

Contrary to much media hype about the baneful effects of service and the "messed up" personalities due to extreme trauma, most veterans become valuable members of society and their community, living normal lives and raising families. They are far more focused than many of their civilian non-veteran counterparts.

Service broadens the human understanding of diverse personalities, ethnic groups, and regional differences of the larger American society. Military people are exposed to Americans of all backgrounds, from the most educated to the marginally educated, from the most affluent to those from the lowest rung of the economic ladder.

THE RESERVE COMPONENT OF THE MILITARY

Many enlistees and officers do not join as full-time regular officers and enlistees in the Army, Navy, Marines, or Air Force. Instead, millions join a Reserve component of the military establishment. Some Reserve *units are fully organized into Divisions, squadrons, wings, etcetera, and can be activated in times of national crisis or emergency. Other reservists are in pools of manpower, without regular drills and training.*

In World War II, for example, General MacArthur had command of 20 divisions after the fall of Bataan and Corregidor in the Philippines. Four of the divisions were regular army, but sixteen divisions were National Guard. These same divisions recaptured New Guinea, the Philippines, and defeated the Japanese on the Island of Okinawa.

Reservists and National Guardsmen have been a vital component of America's military forces from the beginnings of this Nation. In fact, Minutemen, or National Guardsmen, existed before the formation of the regular Army.

Les Allen is a man whose background and experiences illustrate the aforementioned military benefits of national service to the individual as well as the Nation.

LES ALLEN'S BACKGROUND

I was raised in the San Francisco Bay area in the town of Los Altos, an area now known as Silicon Valley. High-tech industries brought employment opportunities for a concentration of engineers, scientists, and technicians to the Valley. Many of the people had advanced degrees in math and science.

My parents were part of that upward-bound Middle Class. My mother was a high school teacher at Los Altos High School, and my father was a Certified Public Accountant. Like most men in my father's age bracket, he had served in World War II. He was a Lieutenant Colonel, and a flight instructor on fighter and bomber planes in the Army Air Corps.

My parents expected me to follow in their footsteps and obtain a college degree in an upward-bound and responsible vocation. I attended Foothill Junior College and earned a two-year degree, but I wasn't really motivated. I was enjoying life, partying, and doing the things young people do. My parents were frustrated with me and finally said, "Enough is enough." They ceased to support my lack of motivation to pursue an advanced degree. It was probably "tough love," because I had to begin thinking in terms of a vocation. At that time the Navy had openings for flight training for students with a two-year college degree. I was accepted for the Navy pilot training program at Pensacola Naval Air Station in Florida. Pensacola is located in the Panhandle of Florida on the Gulf of Mexico. That was the beginning of a career that would last a lifetime as a Navy Officer and civilian administrator. I had a real purpose and direction in life.

TRAINING AT PENSACOLA AIR STATION

Soon I was immersed in navigation, flying, logistics, mechanics, engineering and leadership training. Everything the Navy did or taught had a purpose and ultimate mission. I developed respect for authority and a sense of responsibility for my comrades and the organizational structure. I accepted the challenges and loved the camaraderie with my fellow cadets. Soon I was training on the T-28 and T-34 single-prop aircraft.

The T-28 wasn't just a trainer plane. It was manufactured by North American Aircraft Company, which supplied versions of the plane to many foreign nations for combat operations. It was used extensively in Vietnam. The T-34, the brainchild of Beech Aircraft Company, was designed and built as a trainer for the Navy.

Later I flew the C-45, a Beech-designed aircraft, which was

a dual engine transport/cargo plane, and trainer. Ultimately, I switched to Navy helicopters, learning to fly the H-34.

The H-34 was manufactured by Sikorsky Aviation. It was a piston driven helicopter designed for utility flights and antisubmarine warfare. However, during this period it was used mainly for training by the Navy. My training on that helicopter provided me with a firm foundation for helicopter flying that I needed to be successful in all my future helicopter assignments.

THE U. S. S. RANGER AIRCRAFT CARRIERS

The first *U. S. S. Ranger* of the modern era was specifically designed as an aircraft carrier. During World War II, she was used in the North Atlantic and the Moroccan Invasion against Vichy French forces. After the war, she was scrapped. The next *Ranger* was a Forrestal-class supercarrier and saw extensive service in Vietnam, and later the Gulf War. She was sent to Sasebo Japan during the capture of *U. S. S. Pueblo* by the North Koreans, and later returned to combat operations in Vietnam. During the war in Vietnam, she earned 13 Battle Stars. In 1968, Bob Hope toured the ship and put on one of his famous shows for the officers and men of *Ranger*. After the war in Vietnam, *Ranger* was used for several famous Hollywood movies. In 2014, *Ranger*, like its predecessor, was sold for scrap, a proud ship bought to an ignominious end.

The war was raging in Vietnam when I was assigned to *Ranger*. She was huge, with a crew of 5000 men working in almost every class and Navy skill. The ship was like a small city. There were aviation officers and men, cooks and bakers, finance and payroll departments, propulsion men and officers, aircraft mechanics, maintenance and repair facilities, navigation officers, munitions specialists and handlers, chaplains, medical personnel and medical facilities. You name a need, and personnel were aboard to ensure it was covered. At the apex of the carrier crews were the fighter pilots who took the fight to the enemy over the skies of North and South Vietnam. They were the whole reason for the massive operation.

My mission was as a helicopter pilot assigned to provide search-and-rescue capabilities, utility runs from ship to ship and from ship to shore. The helicopters were deployed as needed in search and rescue of downed pilots at sea or over land. My crew consisted of one co-pilot and two enlisted crewmen who manned the hoists and were trained to jump into the water to rescue downed pilots.

Additionally, sometimes I flew the COD aircraft. COD stands for "carrier onboard delivery." COD aircraft would ferry personnel, mail, supplies, replacement parts, and other high-priority cargo. We flew missions to Japan, Hong Kong, the Philippines, and South Vietnam.

During my two tours in Vietnam, we lost about ten planes as well as pilots. Crashing into the water with the engines at full throttle was particularly dangerous, because it would cause the aircraft to explode, killing the pilots. Sometimes this occurred after a cold catapult launch, when the catapult didn't provide sufficient power to get the aircraft to flying speed.

One event occurred when a jet was landing on the carrier and the arresting gear malfunctioned, causing the aircraft to fly off the angle deck runway with insufficient power to become airborne again. As the plane dropped into the water, the pilots hit the seat ejector switch, but they were too low and too slow for proper chute deployment. Both drowned on impact. We immediately positioned our helicopter over the downed aircraft. We could see the parachutes under the water but we were unable to snag them and hoist them to the surface. Doing so would have caught the shroud lines and chute canopies in our rotors and caused our helicopter to crash.

Eventually, surface boats pulled their bodies from the water.

Carrier operations are full of hazards, with all the volatile fuels, explosives, potential for mechanical failure, and the intense pace of flight activity. To mitigate the risk of serious accidents and incidents, all navy personnel onboard are highly trained and retrained in shipboard safety procedures. One key example of this are firefighting skills.

Fires are probably the biggest danger to shipboard function and life. Fires can occur anywhere on the ship and when they do, the damage can be extremely severe—degrading the ship's operational capabilities quickly. Consequently, all shipboard personnel are required to attend exhaustive firefighting training in specialized schools designed to teach how to use firefighting equipment, put out fires, and prevent conditions that can cause fires. This program is very successful and, as a result, damage from shipboard fires is minimal.

At one point during the war, there were 24 incidents of sabotage in addition to normal accidents and war casualties.

SPOOKING RED CHINA

One operation of our carrier during the Vietnam War was to test the reliability of the Red China defense system. To accomplish this, we entered the Yellow Sea between mainland China and North Korea at night. The carrier turned off all lights, all communication, and other activities that would alert the Chinese military that we were entering their territorial waters. Once inside of that so-called defense area, our carrier turned on all lights, communications, and began all carrier activity including fighter jets doing practice takeoffs and landings. The Chinese defenses did not even know we were on their doorstep until we commenced operations. It must has caused their military real panic being unable to detect a major

United States warship operating so close and yet being undetected. Some Communist military officers may have had some explaining to do to their Communist leadership. The Chinese military was caught exposed to our naval capabilities.

CIVILIAN LIFE AND NAVAL RESERVE SERVICE

After five years of active wartime Navy service, I went back to civilian life with new motivations and ambitions. But the Navy was in my blood, and I continued as an officer in the Navy Reserve, an active reserve component that continued to work closely with the full-time Navy personnel and defense operations.

I returned to college as my parents had wanted years before. I earned my degree in Industrial Engineering at San Jose State University. Armed with a formal education and practical life experience as a Navy pilot, I secured work in the civilian sector in defense-related industries, where I became a manager.

I was married at that time and had three daughters, now all grown, married and successfully employed: one is a nurse, another works in information technology, and one works as a manager for the California Board of Equalization.

NAVY RESERVE DUTIES

Concurrent with civilian employment, I continued my Navy Reserve duties, flying the Sikorsky H-3 helicopter to support the Pacific Fleet's antisubmarine surveillance and flying the RH-53D helicopter (also a Sikorsky aircraft) to support the Navy's minesweeping operations. Both of these activities used fleet-compatible equipment and were designed to be deployed quickly if needed.

In the last assignment, flying the RH-53D, I was the commanding officer of the squadron HELMINERON-19, which consisted of twenty pilots, six aircraft, and approximately three hundred enlisted personnel. Additionally, we had land vehicles, cranes, generators, and small boats. All of this would be needed in the event we deployed.

NOTEWORTHY CAREER EVENTS: SOVIET SUBS WERE IN OUR TERRITORIAL WATERS

We were in training practice coordinated with U. S. Navy submarines off the island of San Clemente near San Diego. Suddenly, the training turned real. We caught the sonar signature of a Soviet submarine in the midst of our training operation. This threat was real and we sent an "active sonar signal" to the Soviet sub that we had him targeted and were aware of his presence. Our helicopters had the capability of firing armed torpedoes. Hearing the "hot sonar" blast, the sub beat a hasty retreat back into international waters. This was during the Cold War era. These incidents occurred from time to time during this period.

Another occurred up in Puget Sound in Washington State when, as before, our sonar detected a Soviet submarine. He was definitely up to no good and was within the territorial waters of Puget Sound, in sight of Seattle and Bremerton. Once again, one of our aircraft beamed a "hot sonar blast" letting him know that he was located and targeted. The sub quickly headed out of the Sound but, in his panic, became entangled in the nets of a hapless American fisherman. The nets were attached to the boat and suddenly, the little fishing boat was being dragged sideways. The fisherman cut his nets to release his boat from being dragged into the open waters of the Pacific. I'm sure that that Russian submarine had a real mess untangling the fishing nets from their propulsion screws.

CIVILIAN ENTERPRISES AND EMPLOYMENT

My civilian employment included management at electronic and optical firms. I worked for Kaiser Aerospace, which was eventually bought out by Rockwell Aerospace Corporation. We built electronic display systems for military aircraft cockpits, including fighters and bombers. Later I was operations manager for Newport Corporation, making high-level optics for military and civilian aircraft.

Essentially, I was employed in two different occupations at the same time, one as a Commander in the Navy Reserve where I was in constant contact with ranking Navy personnel related to that position, and the other as an industrial manager with a demanding boss.

Sometimes there were conflicts between the civilian operations and the Navy operations. One boss once told me to tell my Navy superior who my "real boss" was.

MY ASSESSMENT OF A NAVAL CAREER

From a drifting youth to a responsible naval officer, I had found a career that I truly loved. The action, camaraderie and personal associations, the living conditions, food, and the career opportunities were just what I wanted. Additionally, there was some great shore leave that expanded my travel horizons. Despite the personal benefits, the idea that I was helping to make a real contribution to this nation by my service was deeply satisfying. If given the opportunity, I would do it all over again.

18

FROM UNDOCUMENTED IMMIGRANT TO PATRIOT

The Story of Julio Luna, Career Seabee

Roman armies were more than a fighting force eternally fighting and training. The Roman armies were builders, constructing roads, aqueducts, bridges, baths, and vast coliseums. They left a lasting legacy visible throughout Europe, North Africa, and the Middle East. They built and they fought. They did not laze away in peacetime. During the two hundred years of the Pax Romana or Roman Peace, the Roman soldiers were hard at work building and creating the infrastructure of the empire.

American Seabees are a similar force in the modern age. They build. They fight. They're involved throughout the world working on military, civilian, and infrastructure projects.

Julio Luna's story is post-World War II, Korea, and Vietnam. It occupies a new period in global conflict against Islamic Jihad, a movement rooted in the teachings of Mohammed. It is a movement fraught with the barbarism and brutality of the Dark Ages, a time similar to the Mongolian conquests and the more modern Nazi terror.

Julio's story touches upon many facets of the modern navy and assignments therein. Julio's experiences in the Navy are varied, as he was always open to new challenges.

JULIO'S STORY IN HIS OWN WORDS

I was not born an American citizen. I was born into a family that had emigrated from Spain to Peru almost one hundred years ago. They came to Peru seeking opportunity and a better life. The ultimate better life, however, was not Peru but the United States. My father was an auto body worker in Peru. One year he traveled to the United States on a supposed vacation, a vacation that he used to find employment in New Jersey. He was an illegal undocumented worker, but with his auto body skills and sheet metal experience, he soon found employment in a New Jersey steel mill.

Once established with a job, he sent for his family. For a while I lived with my grandmother in Peru while he was getting settled. I was reunited with my family on August 16, 1978. I was ten years old. I remember the exact date because that was the day Elvis Presley, the King, died. It was the talk of television and the newspapers. Elvis was popular in Peru as well as this country.

I graduated from high school in New Jersey and found employment with Continental Airlines as a translator. I was fluent in English and Spanish. I worked there for 4 ½ years until the recession of 1992. Work was scarce. I toiled at various odd jobs without meaningful or secure employment. I thought I would join the Air Force, but they had a long waiting list as many unemployed youth like me had the same idea. Military service has always attracted youth needing employment. Needing employment, I found the Navy would enlist me almost right away. The advantages of the Air Force and Navy were that they were less onerous than the Army and they could develop civilian-related job skills. The Marines were an option, but, like the Army, they didn't have as many programs for self-development that could be used once a person left service.

SUBTENDER DUTY— THE U. S. S. DIXON

A sub tender is a ship that tends and supports the submarine fleet with fuel, supplies, weaponry, replacement parts and ammunition. My assignment on the ship was working in the engine room on preventative maintenance. That could include mechanical, piping, and electrical.

The ship operated out of San Diego and was eventually scheduled for decommissioning. When the ship was decommissioned and used for naval target practice, I was reassigned to another sub tender in Sardinia, located off the coast of Italy.

INTEGRATION OF WOMEN SAILORS

Back in the Korean War, Black soldiers were integrated into formerly all white units. Previously white and Black soldiers did not serve together. It didn't take long for the integration to be accomplished and opportunities that were unknown in the civilian world opened up for Black men. In my era, the integration was with women. Several commanding officers on my ship were women.

The biology of men and women is obviously different and sex became a big problem, contrary to what the public relations wonks said and the politically correct policies indicated. Those ignored reality. Sex and pregnancy were big issues. If a woman didn't want to be deployed she could become pregnant and be sent stateside. In the early stages of pregnancy, with all the attendant biological issue of bearing children, women could not perform their duties and the burdens shifted to the men. The men felt that wasn't fair. Some men resented women of superior rank they felt weren't qualified but who were promoted because of quotas or even favoritism. Sexual integration created serious morale problems. These male-female issues were observed throughout my twenty years of military service. In my units, up to five females would be pregnant in any given month.

SARDINIA, ITALY

My second sub tender assignment stationed us off a small island 15 minutes off the coast of Sardinia. While there I had the opportunity to interact with the civilian population. On that little island, however, the relationships between the sailors and the residents were not good. Due primarily to drunkenness, there were many fights. I avoided those situations and made friends with a local family. I lived with them for a while and learned to speak Italian. For me, it was relatively easy since Spanish and Italian are both Romance languages. One of my chiefs observed my linguistic skills and told me about an opening at the Naval Language School in Monterey, California. I applied, and for the next six months I perfected my Italian.

ASSIGNMENT TO THE UNITED STATES EMBASSY IN ROME

I served for three years at the Embassy and lived a very comfortable life in a five-star hotel four blocks from there. I wore civilian clothes and got to meet some very famous people. Two of the nicest politicians were Senator Joe Lieberman and Senator John McCain. They treated us with great respect and kindness. They even took us to dinner. Hilary Clinton was friendly and pleasant, but Dick Cheney was a @#$@!%&#. Sorry but I can't say in print the kind of man he was. I even met President Jacques Chirac from France. He had quite a reputation as a "Ladies' Man."

POPE JOHN PAUL II

I am Roman Catholic, and being stationed close to the Vatican, I attended Mass there when the Pope was the officiant. Pope John Paul II was a linguist and gave the blessing in 15 different languages.

I saw him more than once when he was riding in his Popemobile. He would bless the crowds and pray for all people, Catholic or otherwise. I knew that he almost died from a Turkish Muslim's assassination attempt. His stomach had been riddled with bullets. Ever after he was frail and in poor health but he continued to do his duty as Pope. He even visited his would-be assassin in prison and forgave him.

The Pope was so frail and trembling yet such a good man that I cried when I saw how he was suffering. Many of the onlookers whether Catholic or not cried as well. I have always had great respect for this Pope.

A NEW ASSIGNMENT- MILITARY POLICE

I enjoy changes in job assignments and experiences. I left the American Embassy in 1999 and was assigned as a translator for the 5th Fleet Commander in Naples, Italy. That was followed by training at the Military Police School in San Antonio, Texas.

From the wholesome and upscale life as a translator at the American Embassy, I was thrust into the seamy side of life. Assigned to a base near Naples, we had the duty of checking on the welfare and well being of Navy personnel living off base.

Naples was Mafiosa territory and occasionally, we would see dead bodies outside the base. One day an Italian man was parked outside the base when a hit man on a moped came by and shot the car's occupant in the head. Another time, one of the Navy Chiefs was involved in some kind of "black market" operation. His body was found floating in Naples Bay. This was followed by another incident involving the Italians. Two kids went to the wrong house and knocked on the door. They were both shot dead. There were incidents where American sailors became involved in the drug trade. They were arrested by the Italian police and confined to the base while awaiting trial. Justice is slow moving in Italy and these suspected drug dealers were still confined to the base when I was reassigned. I never found out what happened to them. However, when crimes were committed by Navy personnel and they were convicted, they were sent to the military brig in Germany.

As Military Police we had to deal with crime, corruption, traffic accidents, deaths and other law enforcement issues. It was a very different experience from being a salt-water sailor or an interpreter.

KOSOVO- PROTECTING CIVILIANS FROM MASSACRES

The Balkan countries of Europe had been under the control and domination of the Muslim Ottoman Turkish Empire for hundreds of years after the Turks invaded Europe from the outer edges of Mongolia. There was always much turmoil under their harsh Islamic rule. Most of the Ottoman/Turkish subjects in the Balkans were Orthodox Christians who were allowed their religion as long as they paid a special tax to the Sultan's government and accepted the extreme regulations imposed against Christians. Some of the Balkan people converted to Islam because they then would be treated better than as Christian subjects.

By the beginning of World War I, the great Austro-Hungarian Empire was butted up against the Ottomans/Turks. As most people know, World War I began in the Balkans with the assassination of the Archduke Ferdinand and his wife by a Serbian nationalist. Subsequent to World War I, the Ottomans/Turks lost much of their formerly conquered territories. A new nation was created out of several of the former Ottoman territories. It was a hodgepodge of different nationalities and religions, a difficult scenario for any country and its rulers. The new nation was Yugoslavia, and after World War II, a communist dictator ruled the country with an iron fist. The quarreling factions were kept at bay and internal peace was imposed on the citizenry—there is something to be said for dictatorship such as imposed by Tito. By force and might, he kept the people from killing each other. But Tito died, and the next government was unable to maintain peace and order. The hostile, quarreling segments of Yugoslavian society broke out in a ruthless civil war, where atrocities and genocide were all too common.

The history of these conflicts is too convoluted to discuss, but it was the result of Ottoman rule that left a large section of the country Muslim in the midst of Orthodox Christians who had experienced harsh Muslim rule, enslavement, and abuse.

When we were sent to Kosovo, it was to prevent a massacre of the population by Serbs and the Serbian military. I spent six months there. I earned a combat action ribbon since there was occasional hostile mortar fire in the area. The people of Kosovo loved us because we were their shield against annihilation by the warring factions. I felt that the American forces helped to stabilize the region and prevented further civil and military atrocities. I felt good about what our forces did in Kosovo. We saved the lives of women, children, and families.

I WANT TO BE A SEABEE

While in Kosovo, I observed the activities of the Seabees. It seemed like a positive change from some of the Military Police duties. Policemen do not always see the best side of humanity.

The Seabees were an impressive group with a sense of energy and esprit de corps. They fought and they built. I thought that was a neat assignment and I was yet again ready for a change. By 2001, I was ensconced in MCB40 at Port Hueneme, California, the West Coast home of the Seabee battalions.

My new assignments involved positive and enjoyable activities where I could see the impact of Seabee work. We went to Spain where we built housing, churches, and schools. Later we went to Guam where we helped clean up after Typhoon Chataan. Ironically, Chataan means "rainy day." The rain and winds were extremely destructive to Guam, with almost 2000 homes destroyed, fields and towns flooded and the island infrastructure severely damaged. Our Seabees rebuilt roads, removed downed trees, and restored electric service and potable water supply. Guam is an American island, and the people loved us for all the work and help we gave them. I was awarded the Navy Service Medal for my work on Guam.

While a Seabee, I became a plumber. In 2004, the Gulf Coast states in the United States were hard hit by super hurricane Katrina. (In the Western Hemisphere these powerful storms are called hurricanes. In the Far East, they are called typhoons.) We were sent to Gulfport, Mississippi, where the storm surge had destroyed most of the buildings along the shore. The storm surge had raced as far as two miles inland. Major shopping centers were destroyed, along with cafes and other businesses.

We built temporary housing for the people, installed showers, cleared roads and did whatever was needed to restore the community to normality. The people were so happy to have the Seabees there. One Black man told me it was the most help he had ever received. I felt good about this assignment because we were helping our own people, our fellow Americans. We rebuilt schools, churches, homes for veterans, shelter housing for FEMA, and facilities for the Red Cross. I received my third Achievement Medal Award for the work I did in Gulfport that summer.

There was one negative in Gulfport: the local contractors were angry with the Seabees because they said we were stealing their jobs. Though we felt good about helping American people in desperate need, many of whom had lost everything, another downside was the miserable summer heat and humidity in Gulfport. But in our hearts and minds, we knew, as American Seabees, we had done a good thing for our people.

One of my last assignments was at the Sigonella Base in Sicily where I worked in plumbing. I enjoyed the friendly Italian people.

My twilight combat tour with MCB 5 was to the war in Afghanistan.

AFGHANISTAN

We were flown by commercial airlines to Kuwait on our first leg of our flight to Afghanistan, half way around the world. The total distance was 7253 miles from the USA. Then we flew to Helmond Province and Camp Leatherneck. This was in southwest Afghanistan and in the heartland of the Taliban, who were quite active in the area. Nearby was a British base called Camp Bastion. It was a totally different experience from my previous military assignments. We had to be on guard against any danger from unknown personnel. The Afghan people did not like us. They hated our female military staff, especially since women were regarded so poorly in primitive Muslim societies. We never knew who was really a friend or foe. Even the Afghan military personnel we trained were known to be a potential threat. I knew that Afghanistan was one of those faraway places where wars have been fought by the Mongols, the Muslims, the English, and the Russians. No one was ever able to subdue these primitive and warlike people.

OUR MISSION NOT DEFINED EXCEPT AS BUILDERS

Unlike World War II or any of our previous wars, the rationale for our American presence in Afghanistan was not clearly defined. Our mission was to build offices, housing pods with air conditioning, concrete pads, and huge warehouses. The daytime heat was unbearable, reaching 120 degrees and above. We did not work in the day. We worked at night. About once a week rockets would be fired from the base whenever enemy activity was detected or Taliban were observed trying to infiltrate the concentric defenses of the base.

IEDS—UNCONVENTIONAL WARFARE

Unlike the great wars in the past where armies faced and fought each other in readily observable positions and uniforms, the war in Afghanistan was fought through stealth, cunning, and improvised explosive devices detonated from well-hidden places, followed by an ambush. Warfare in Afghanistan is totally unconventional. Periodically we had to make trips of eighty miles to Kandahar for supplies and to drop off equipment. The convoys included up to 18 trucks heavily armed and escorted by Marines with 50 caliber machine guns and M-16 rifles. During my tour in Afghanistan, the Humvees had been strengthened so they were more blast proof. We had confidence in the Marines' ability to protect the convoy. Ironically, Seabees and Marines are Navy personnel, but we were far from the ocean, stationed in the middle of a God-forsaken desert. One day while in convoy with 15 Humvees, one was hit by an IED. Fortunately no one was hurt. Immediately, the Marines and Seabees set up a perimeter of defense. On this particular trip, we only had five Marines so we returned to Camp Leatherneck for more support.

NAVY BASE AND AIR FIELD IN ROTO, SPAIN

My next assignment was in Roto, Spain, where we worked on upgrading military installations and helping the local people building churches and schools. Roto, Spain is in the Southwest corner of Spain and houses a huge American military complex with a naval port and airfield. It's close to the Straits of Gibraltar and a halfway point between the United States and the Middle East.

The Spanish people were friendly and I enjoyed mixing with them since I spoke Spanish and was of a Spanish heritage. I liked it best when as Seabees we were good-will ambassadors helping other people.

HOME TO PORT HUENEME PENDING RETIREMENT

My final year with the Navy and Seabees was at Port Hueneme in Ventura County, California. I spent the remainder of my enlistment and career as a range coach at the nearby Point Mugu Naval Air Station. I trained Navy personnel on the M-16 Rifle, the Mossburg Shotgun, and the 9mm Berretta pistol.

FAMILY LIFE AND "RETIREMENT"

I was married while in the service to my wife Debbie. We have one nineteen-year-old son, Jack. Shortly before retiring, we moved to the Ojai Valley in a semi-rural area where my wife can enjoy horseback riding. However, as a young man in my forties, I can't afford to live on my Navy pension, so I've gone back to work. Of all places, I was employed at my old Seabee Base at Port Hueneme as manager of the plumbing shop. My Seabee training in that field qualified me for the new position. So, I remain connected to the Seabees, a proud branch of Naval service.

Prior to my final enlistment, I joined the Ojai Veterans of Foreign Wars Post 11461. I became quite active, working to advance the cause and understanding of what it means to serve in the Armed Forces of the United States. I have given talks on service at the Valley-wide Memorial Day events, and have assisted in meeting the public, and in organizing the promotion of Buddy Poppies for widows and disabled veterans. I participate in the Annual Christmas Dinner sponsored by the Post to remember the widows of veterans and families of active duty military

CONCLUSION

Unlike the old-time veterans from World War II and Korea, all in their 80s and 90s, and the Veterans of Vietnam now approaching their 70s, I'm still a young man and have many years of service left in the workplace and in my Ojai community. I hope to be known in this community as one who makes a positive contribution to our society, as have the veterans of our past wars.

THOMAS ROSS—A SERVICE BRAT AT WAR IN AFGHANISTAN

As written by Seabee Thomas Ross

EARLY LIFE IN A NAVY FAMILY—CONSTANTLY MOVING

I grew up in a military family and was born in Bremerton, Washington. We didn't live there long before moving to California where we stayed until I was in the 2nd grade. We then moved to Hawaii (Oahu) where we lived until I was in the beginning of the 5th grade. My dad then got orders to go back to Washington State during the middle of my school year, and I started 5th grade in Port Orchard, Washington. My parents bought a house and I finished my schooling there and graduated from south Kitsap high school in 2004. I joined the military in 2004 to follow my father's and grandfather's footsteps. Also just to get away from the town I grew up in.

I CONTINUE THE MILITARY SERVICE TRADITION

Sent off to Illinois for boot camp, I remember getting off the bus and getting screamed at right away. I went through boot camp and got sent to Gulfport Mississippi for my builder A school. I finished that after a long 12 weeks of training to learn my job and got stationed in Port Hueneme, California with Naval Mobile Construction Battalion Five.

FIRST DEPLOYMENT—GUAM

I then went on my first deployment in 2005 to Guam. I had no idea what to expect. It was a green deployment, meaning not the desert. No weapons! I got there and, having just turned 18, the first thing I was told was, "HEY YOU CAN DRINK HERE." Great first impression of the Seabees! Deployment went on and we worked our butts off. I got to experience the culture of Guam. Family was a big thing over there. "POT LUCK" ON THE BEACHES! The most amazing food I had ever had. Everyone was great. MMA (mixed martial arts) was also a big thing over there. We got to the end of the deployment and then hurricane Katrina hit Mississippi, which is where the battalion was stationed that was going to relieve us. We spent another month and a half in Guam, where you can do everything doable in two weeks. JOY!

We finished deployment and went home. Homeport is train, train, train!

SECOND DEPLOYMENT—ALL PASSED OUT FROM THE HEAT

After a 10-month homeport, I got sent to Djibouti, Africa, during the time when Somalia was getting a little hectic in the world. As before, I had no idea what to expect. This time we were getting hazard pay and all that fun stuff and carried weapons and wore full battle gear. For a 20-year-old, this was pretty nerve-wracking, but exciting! We flew into Saudi Arabia on a C-17 for some reason. Could be refueling? Who knows? We got off the plane and I about died! It was 140 degrees outside! We rushed into the terminal and waited for word. After about six hours of waiting, we put our full battle rattle on (about 70 lbs) and headed out to the tarmac (runway). We climbed into the C-17 in full gear. Once you are in your seat, there's no getting up. You're stuck! Got to pee? Just let it go. So we're sitting there and it's hotter than outside. We asked the flight crew to turn the AC on and they said not until we were in the air. We sat there for 2 hours in probably 180 degrees. That's all I remember before waking up when we were airborne. I looked at the guys next to me and everyone was slowly waking up. I was confused, and then someone said we all passed out from the heat. We had 96 people in that plane! EVERYONE PASSED OUT! SERIOUSLY? That can't be how business is supposed to operate in the AIR FORCE is it??

SHOCKED AT THIRD-WORLD COUNTRY IN AFRICA

After that nightmare of a plane ride, we landed in Africa, and I got a shell shock of what a 3rd world country really was. For all those in the United states who think life sucks or we have it so bad, by all means head to a 3rd world country and see how life really sucks. Everything you take for granted is gone. Traffic lanes? What is that? People drive all over and at high speeds. Trash is everywhere, their goats are eating the trash on the ground and their kids over there look like they have not have a bite of food in weeks.

Everywhere we went there was TPI (TWO PERSON INTEGRITY) meaning if you left the base, you had a buddy with you. Even though it wasn't a war zone, there were still people out there trying to hurt you. I was only able to leave the base one time due to security reasons. I left with about four other guys and we took a cab out to the city. Scariest cab ride of my life up to this point. Going about 80 mph, cutting through lanes, missing cars and people by inches. We get to the city and decide to go look for souvenirs to buy to bring home. Word of advice: remember you're in a 3rd world country. If you see a Rolex for sale and its 20 bucks, it's probably not a Rolex. We found all kinds of little trinkets and whatnot. The best part about shopping there: bartering!

There were no set prices or stores. Everything was outside in little huts or tents. A guy could say he wants 50 bucks for a wooden elephant and you turn around and say, "Hey I'll give you ten bucks," and trade him something from America and he'll jump on it! They love American money over there, so as soon as you flashed it people would swarm you, trying to get you to buy what they were selling. You had to be real careful though, because of the pickpockets there.

After a solid few hours of shopping, we strolled around some more. I'll never forget, I saw a man walking a hyena on a leash. Blew my mind! We ended up in a bar later in the evening to have a few drinks, and it was crazy. Every bar you stepped in was like a club: disco lights, strobes, and loud music. We didn't like the environment after a while and decided to go back to base.

We had multiple projects to work on around the base. One project was in a spot outside the wire called Ball Balla. We built a schoolhouse for the local community. We had a few locals who helped us with communication around the area when we went out for materials and whatnot. The locals also provided some security.

One day we had a kid, no more than 12 years old, take a nail punch (looks almost like an ice pick) from our job site. One of the security guys took off after this kid to retrieve the item. We tried to tell the guard it was fine, let it go, but he kept going. About two hours later, the guard comes back with the tool covered in blood and tries to return it to us. Needless to say, we told him to keep it. WE NEVER ASKED WHAT HAPPENED! We closed up work and left. The next morning we came back to feces smeared all over our door handles and locks. REALLY? We were trying to help, but they didn't care.

We finished the building, had a ribbon cutting ceremony, and left. As soon as we got to base, a local man called and said that the building had been torn down and stripped for material for people's little shack huts that they live in. Talk about a slap in the face! All that hard work for nothing!

ETHIOPIA AND KENYA, THE CONTRAST FROM DJIBOUTI

We had trips to Kenya and Ethiopia where we saw a culture other than that in Djibouti. We sang, danced, drank and had a blast!

Building tents for locals to live in and all sorts of good community work was most enjoyable

SEYCHELLES—PARADISE IN COMPARISON

The Seychelles was a paradise in contrast to previous assignments. Pictured is the coastline of one of the islands. Most people never heard of this island nation. It's located off the east coast of Africa in the Indian Ocean.

I then got pulled back to be sent out on a detachment with 12 other people to an island nation called Seychelles. If you have ever seen the movie *Cast Away*, that beach is on Seychelles. IT WAS PARADISE! We went from a disgusting, trashed, hot, stinky place to a tropical paradise, and I was not complaining at all.

We got there and were tasked with remodeling a schoolhouse and building a brand new mortuary. By the way, our projects were right on the beach. We were very popular on this island, being the only Americans. We met all the locals and also the rich locals on the island. GREAT PEOPLE! We met a British guy named Peter. He owned the biggest and most successful resort on the island. Needless to say, we got spoiled. We partied with him and his family, went to classy dinners, boat trips, and got full access to his swimming pool, which had a swim-up bar. I felt like a celebrity. Oh, and to top it off, we ended up on TV because of the work we did, and, of course, Peter's connections.

I met the country's President, smoked a very expensive cigar, and honestly lived like a god. We did work our butts off over there though. We finished the school and the mortuary and even had a plaque made by the locals with our names on it and embedded into the mortuary in the front for everyone to remember us. What an experience to have at such a young age.

We traveled to many of the islands via boat. The first trip I took on a boat, the captain stopped the boat and told everyone to shut up and not to move. I was very confused. And then I saw a shadow in the water going under out boat that looked to be as wide as 3 school buses and as long as 6 school buses. It was a huge whale shark! The boat driver didn't want its tail coming out of the water and smashing us. It passed us and then came out of the water about 20 feet from us. AMAZING! Just another thing about this place that blew my mind. This made me want to go fishing. We set up a trip and went fishing for hours. So much fun! Caught huge tuna and barracuda. And then someone caught a six foot long shark! The driver stabbed the shark to death, tied it to the side of the boat, and went back to shore. No more fishing he said. This shark will feed hundreds of people. We brought the shark to shore. He started chopping it up and people were coming by and buying meat from him. I guess we hooked him up, ha-ha.

LIQUID TURKEY FOR THANKSGIVING—WOW!

It was thanksgiving and the island didn't celebrate it, so we talked to Peter about trying to get a turkey imported to us. He said, "I got this," and he took us out to this super secluded island that he owns (the one where Prince George and his wife Kate went) and we had a Thanksgiving party. Well there was no turkey. But he did bring us a bottle of Wild Turkey 101. Ha ha ha, I guess he got us a turkey! After the fun activities, I decided to go snorkeling. The water was so clear you could see 50 feet down. I snorkeled for about 20 minutes and a giant shark swam right in front of me! Words were screaming, I'll leave it at that. I rushed my butt back to shore and never went back in. We enjoyed the island and the Creole food they had, and then we had to pack up to go back home. Deployment was over "SAD FACE ☹." The Air Force had to pick us up on the main side of Mahi. Small airport for small airplanes. The air force was bringing a C-130! Not so small. This plane barely made the landing without crashing. Now we got to take off? UGH! The pilot backed up as far as possible and gunned it. We probably took out a fence at the end, who knows? It was close, though.

HOMEPORT AGAIN—TRAIN, TRAIN, TRAIN

Homeport it was! Training training training. Field exercises at Fort Hunter Liggett and more training. I was a 3rd class petty officer by then. We stayed in homeport for 12 months this time. Then it was time to deploy again. Number 3! Okinawa Japan.

DEPLOYMENT #3—OKINAWA, PHILIPPINES, THAILAND

Off we went. Another new culture for me to explore, and explore I did. Food testing, sightseeing, and learning the way of life in Japan. It was very interesting over there. I loved the food! I saw the world's 2nd largest aquarium and went on hikes to waterfalls and all sorts of outdoor activities. We worked on a few projects for about three months and then I went on another detachment to the Philippines and Thailand. I wasn't able to go out in the countryside. Too much threat there, but we built schools, met locals and then left for Thailand.

Thailand was a different story. We were able to go out. To start we were on the lower side of Thailand, about 4-6 hours south of Bangkok. We stayed in an abandoned building on cots (little fold up beds). Bugs were so bad there, we had to put up netting around our rack and in the morning you would have to knock them off just to get out of bed. I had a bat on my net once just hanging there. Yikes. We were building a school for the kids here. HUMITITY like no other! SOOO HOT! We got really ahead on our project and my chief at the time decided to let us all go up to PUKET. (foo-ket). We stayed at a five-star hotel for 50 bucks a night! NICE! Thailand was very clean, and very similar to the United States. They had a subway and a few American places, which made it feel like home, but I wanted a change. We ate at all the local hotspots. If you don't like spicy food, Thailand probably isn't your best place to visit! I went parasailing, jet skiing, and hang gliding. We sat on the beach and had people trying to sell us trinkets. We had a lady come by with an aloe plant and rubbed it on our skin to protect us from the sun. It felt amazing. We then decided to get hentai tattoos. WHY? I have no idea.

Nightlife here was something else. All I have to say about the nightlife is adults only. Don't bring your kids. After everything, we finished out projects and went home for another homeport for 12 months. By the way, most of my deployments were 6-8 months. Homeport was train train train again.

AGHANISTAN, THE WAR ZONE—4TH DEPLOYMENT

*Seabee Battalion Arriving in Afghanistan—
Unbelievable desolation*

I was ready for my 4th deployment. AFGHANISTAN! Not the most thrilling thing to hear after everything else I'd done. But at the same time I was pretty pumped to go. I wanted to be a part of the "war on terror" and see what it was all about. Boy did my attitude change when I got there! There are going to be a lot of things I leave out due to security reasons and things I just can't talk about, so I'll tell you as much as I can.

At the beginning I was with a 10-man crew in J-BAD (JALALABAD) with the team guys (SEAL TEAM). We were tasked with building a large 100' x 60' K-span building with a full interior (offices, mechanical shop, and work stations). As you know, Afghanistan is a war zone and weapons are carried.

Since we had the SEAL team with us we weren't required to carry weapons, nor have any patches or identifying marks on our uniforms. It was pretty nice not having to worry about that kind of stuff and just focus on the mission. My 2nd day there a car bomb went off at the front gate, killing multiple people (both Navy and civilians). This was my eye opener: I was in a place where anything could happen at any time. Then it seemed that on a daily basis mortars were being launched at us. They didn't hit anywhere near, but they were around and they did shake the ground when they hit. At first it was nerve-wracking and kept you on edge throughout the day. But after a week of it you tended to ignore it and just keep hoping their aim sucked.

After about a month, we finished out project. A lot of things went on throughout that month with the team and other people that I can't write about, but these events really opened my eyes to what war really is. So since we finished we were sent away for more tasking. This time I got hand selected to join a team in Kandahar with three other people. We were strictly team support. I was pumped! It was a chance to get away from the Seabees, see the other side of things and be a part of the overall big mission picture.

We got there, got read in, and were briefed on where we were and the threat around this area. Seemed like this place was a hot target for the enemy. So I met up with the people I worked with and got to know all the key leadership guys and what they expected of my guys and me. We did do Seabee work for them if they needed something, from satellite platforms built to weapon staging areas and so on. The coolest part of meeting people was that it wasn't just the Navy team guys here. I met the recon guys, the Rangers, the Delta team, and so on. Very interesting bunch of guys. Most of them stuck to themselves and their team for their own reasons, but some of them were very friendly and liked to get to know you and where you were from. I got to know the recon guys very well. They asked if we could do some work in their tent for them. (We lived in 40 by 20 open bay tents on little twin-sized beds.) They asked us to build them some privacy walls and shelves and a game area to relax in. Of course this wasn't my tasking, so I made a deal. (In the desert, it's all about networking. Give something, get something). My guys and me wanted a TV and a game system of some sort with a couch. They said, "We'll see." Of course, I thought there was no way we'd get our stuff, but I decided to do the work for them anyway.

I finished up and came back to my tent and there it was. A 50-inch TV, leather couch, and a Nintendo 64 with 4 controllers and Mario cart! Ha-ha! Talk about a huge morale booster! I ended up hanging out with those guys a lot—learning the ins and outs of what they do and of course playing Call of Duty on their Xbox every chance I got ☺.

Throughout the days mortars hit as they did at the last FOB (forward operation base) but here they hit close, sometimes, though rarely, on base. And these shook the ground big time. Once something like that happens, you run to a bunker and wait for an all clear. Sometimes it was a little scary but everyone was in the same situation and we just told stories and laughed to forget about it.

One of the most nerve-wracking moments there occurred when I was in the shower after the gym. Our showers were in a conex box. (It was a shower, that's all that mattered.) I was in the shower and a HUGE explosion went off, knocking me out of the shower and onto the ground. I sprinted to the closest bunker, butt naked. I get to the bunker and everyone just laughed. But at the same time you have to do what you have to do. We came to find out it was a massive car bomb at the front gate, which killed a lot of people. So it was a huge shock.

Time went on and I met some of the snipers there. Made a few targets and whatnot for them. Got to know them pretty well.

ASKED TO BUILD A MEMORIAL

One day I got woken up by a team member asking me to make a memorial step box. (Black box with two steps for boots, Kevlar helmet, and your rifle to rest on.) I said ok, no problem—will make it. I asked him what happened. That sniper who I had been getting to know had been shot. He lost too much blood, and had passed away that night. It hit me really hard. That was the guy I'd last talked to before I went to sleep that night.

We had a memorial for him later that day and everyone got in their uniforms for the first time since I'd been there. It was really different seeing everyone in their uniform with rank and names and whatnot. Over there, people wore their own clothes or whatever with no names or anything and called each other by their first names.

There were some other major events that took place that deployment that I can't talk about but really made me think about life and how lucky I am just to be alive and walking.

Thomas Ross, right, receiving award

FOB—CLOSER TO THE ACTION

Later in the week I got tasked with going with the team to a small FOB (Forward Observation Post). I can't tell you the name but it wasn't too big and it was in the middle of nowhere.

They needed me to build two tent decks 100' x 40', install the tents and then build them a staging area for all their weapons and gear for missions. This was the first time I'd ever ridden in a Blackhawk helicopter. Pretty cool! We got to this FOB, met up with the EOD team there and got the location of work. There were other NATO forces on this base as well. British, French, and Romanians. I got to shoot a few weapons from the other forces, which was really fun. AK47s, MP5s, P90s, just to name a few. Needless to say I had fun.

An airplane takes troops to the Forward Observation Post area in Afghanistan. Notice the dry, desolate landscape below as Thomas Ross mans his weapon.

THE KID ON THE BIKE EXPLODED

Later that evening we had a kid on a motorbike riding the perimeter of our camp. I'm not too sure of the situation but things got serious and they were trying to get him out of there and I guess he wouldn't listen and came closer. They opened up on him. The bike blew up and left a huge crater. I guess it was a bomb. Anyway, I finished up the work they wanted after about a week, and we took the Blackhawks back. The very next day, I heard that the camp we were at got hit really hard and a lot of people lost their lives. Shell-shocked!

UNBEARABLE DAYTIME HEAT

We got back, and boy was the heat picking up. We no longer worked days because it was just too hot. Night shifts were authorized! Plus that's when the teams worked anyway, which helped us get tasking that much faster. We worked a lot of nights getting a lot done and I was up for my Seabee Combat Warfare pin. I had to go back to Bagram, where my higher leadership was currently stationed. I got back, completed my board for my pin and a mortar hit 200 feet from where I was standing. It didn't go off! But it did go through the guard shack. The roof of this guard shack was sheet metal, and it cut the two guards to pieces. Talk about a messed up deployment. I understand how people get PTSD. Thank God, I don't have trouble with it. After all that I ended up back in Kandahar.

Heavy Machine Guns Firing in Afghanistan

I DIDN'T KNOW MY NAME

I worked about another month and was getting ready to go home. I had one day left on deployment! JUST ONE! And then it happened. I woke up strapped to a stretcher being loaded on a C-17 on my way to Kuwait with my head split open and I didn't even know my name. I can't tell you the backstory to what all happened, and honestly I don't remember. But I got to Kuwait and was wheelchaired into the hospital there. They stuck me with about six different needles in my arms and asked me questions about who I was, where I was, and my name. I didn't know where I was, but I did remember my name this time. I was really confused at that point. They told me to stick my arm out and touch my nose. So I did, except I couldn't touch my nose. As hard as I tried I couldn't do it. And I knew something was wrong.

I stayed there for about a week and they said I might have neurological damage. Yes, I was a little scared at that point. I then got strapped back in a stretcher on a C-17 and was being sent to Germany. I was pissed because I was supposed to be home off deployment.

HOSPITAL AND THE CARNAGE OF WAR

When I got on the C-17 there were multiple guys there. Some had been shot, some who'd hit IEDs, and whatnot. We all got to Germany, checked into our rooms, and met our doctors. Those who couldn't walk were escorted, those who could were given a schedule for appointments and told to be at your appointments and then do what you want. I was given $200 in cash to buy clothes since mine were cut off, and I was let go. I met my roommate. He was an Army guy with over 100 stitches down the center of his face. I guess he was on a 15-foot high tower watch in Iraq and when his relief came up to relieve him, he went to help him up and was pulled off the tower and fell 15 feet on his face. Talk about bad luck.

I met more guys from the desert, some blown up and some shot. One guy had a heart attack, some had undergone seizures. I went on with my days going to appointments. I was able to touch my nose at this time and everything seemed like I was back to being me again. So of course being in Germany, where my dad was born, I asked if I could go off the base to check it out! YEP! So I left with a friend I met and we went to explore. It was amazing. Went on the autobahn and tried a couple beers out.

I was in Germany for over a month before I was able to come

home. They flew me from Germany to San Diego, where I stayed for a week for MRIs and all sorts of tests. I was so mad at this time because I was so close to being home but couldn't go home. I finally got released and I went home. YAY! I had about three months left in NMCB 5 before I transferred.

Thomas Ross being presented as Sailor of the Year

I MEET MY WIFE WHILE ON LEAVE

I'd been gone for so long I decided to go hang out with friends at a house and that's where I met Elise, the woman who is now my wife. Long story short, she was going up to Washington State for school, I was getting stationed up there so we continued seeing each other. We got married three years ago in Bremerton, Washington. From Washington my orders sent me back to Port Hueneme before I was deployed to Atsugi, Japan. Back to Port Hueneme for several months before being redeployed until orders were over and back to NMCB 5 I went. I was sent to San Clemente Island for my 6th deployment. Shortly before my 6th deployment to San Clemente, my daughter was born at the Community Hospital in Ventura.

THE FUTURE AS A CAREER SEABEE

Military service can be a pain or a pleasure. Seabees in particular are in a position to see the world and experience different cultures both good and bad. It's a learning experience that transcends any routine employment of the stay-at-home civilian youth. As a career Seabee, Thomas will have ten more years of service before retirement to the civilian world. The complete saga of Thomas Ross' life and service is still to be written.

Seabees at Work- Thomas Ross in Foreground

One aspect of military service affects home life, with months away from wife and family. It creates special hardships in family relations, and there are aspects of watching children grow and mature that can be problematical. It means that military wives have to be more self-directed and self-reliant as they care for home, children, and their assignment in the workplace. With the all-volunteer military, it means that spouses are under constant pressure to respond to the military needs of this country. One unique feature of the Seabees is that they're a very close knit and supportive military community, especially those who make it a career.

Author David Pressey

Born in Maryland and moved to Los Angeles while still an infant, David Pressey is a descendant of American veterans from the French and Indian wars, the Revolutionary War at the battles of Lexington and Concord, the Union Army at Vicksburg, and the Spanish American War. He joined the National Guard at age 17 and served in Korea and Japan. Dave remained in the Guard until 1967. After the Korean War, he went to college and received his masters (magna cum laude) from USC. Throughout a long teaching career he earned additional degrees and credentials. He married his bride Elizabeth at St. James Armenian Apostolic Church in Los Angeles 60 years ago. Elizabeth is a first generation American from the Christian refugees fleeing Islamic massacres in Turkey. He became an Ordained Deacon and later priest in the Anglican Church. Dave often presides over ceremonials and funerals for veterans. He is a charter member of the Ojai, California VFW Post and along with fellow vet Chuck Bennett created the Veterans' Stories Series of books in 2000. He is the sole editor and author of Volume III. Dave and Elizabeth have traveled extensively through Europe, Asia, North Africa and North America. They reside in Ojai, surrounded by their children, grandchildren and two great grandchildren.

BIBLIOGRAPHY

The Holy Bible, King James Edition

The Books of Moses

The Torah

Alvarez Jr., Everett. *Code of Conduct* New York, New York: Donald I. Fine, Inc., 1991 (235 pages)

Becker, Carl and Sidney Painter, Bert Loewenberg, Yu-Shan, James Blakemore. *The Past that Lives Today* Morristown, New Jersey: Silver Burdett Company, 1952 (854 pages)

Brebitsky, William. *A Very Long Weekend, The Army National Guard in Korea* Shippensburg, PA 17257, Whitte Mane Publishing Co. Inc.,1996 (293 pages)

Carey, John. *Eye Witness to History* Cambridge, Massachusetts: Harvard University Press, 1988 (706 pages)

Delk, James D. *The Fighting Fortieth, In War and Peace* Palm Springs, California, 1998 (424 pages)

Goldwin, Robert, ed. and Robert Lerner and Gerald Stourzh. *Readings in American Foreign Policy* New York, New York: Oxford University Press, 1959 (496 pages)

Goulden, James. *Korea, The Untold Story* New York, New York Times Books, 1982 (690 pages)

Gurney, Gene. *A Pictorial History of the United States Army* New York, New York: Crown Publishers, Inc., 1966 (815 pages)

Hill, Jim Dan. *The Minuteman in Peace and War, A History of the National Guard* Harrisburg, Pennsylvania: The Stackpole Company, 1964

Hirsch, Joseph C. *Pattern Of Conquest* Garden City, New York: Double Day, Doran and Co., Inc., 1941 (309 pages)

McCartney, William F. *The Jungleers, A History of the 41st Infantry Division* Washington Infantry Journal Press, 1948 (208 Pages)

Morison, Samuel Eliot. *The Oxford History of the American People* New York, New York: Oxford Press, 1965 (1153 Pages)

Pratt, John W. *A History of United States Foreign Policy* Englewood Cliffs, New Jersey: Prentice Hall, Inc., 1960 (808 Pages)

Van Creveld, Martin. *Technology and War* New York, New York, The Free Press, A Division of Macmillan, Inc., 1989 (342 pages)